Building Research Establishment Report





# Structural damage in buildings caused by gaseous explosions and other accidental loadings, 1971-1977

**R J Mainstone,** DEng, CEng, MICE, FIStructE, FRSA, FSA, HonFRIBA
Building Research Station

**H G Nicholson,** BCom
Construction Industry Research and Information Association

and

**S J Alexander,** MA, CEng, MICE, MIStructE
During the period covered by this report, of Ove Arup & Partners and Kenchington, Little & Partners; now of T H Engineering Services Ltd

Department of the Environment
Building Research Establishment

London, Her Majesty's Stationery Office

## SUMMARY

A survey of accidental structural damage to non-industrial buildings throughout Great Britain was started in April 1971 to provide more reliable data than were then available for the framing of design recommendations and regulations. Press reports have been the primary source of data, supplemented by selective inspections of damaged buildings. For the first two years the survey was carried out by CIRIA under contract to the DOE, and it has been continued subsequently under contract to and in association with the BRE.

The present report covers the whole of the first six years of the survey and supersedes an earlier report on the first two years. It presents tabular data on the frequencies of occurrence of accidental loads of different severities on different types of non-industrial building and deduces average annual risks and, in the case of gas explosions, risks in terms of energy consumption. Damage due to deliberate bombings is not considered in detail. Accidental explosions were found to be the principal cause of damage, with impacts by road vehicles second in importance.

It is concluded that present regulations are not unreasonable if the cost of meeting them is considered reasonable for the avoidance of about one serious partial collapse of a building of five or more storeys every 25 years. The present annual cost to the country of the damage that is caused is estimated at about £1½ million, with between 80 and 90 per cent of this attributable to accidental explosions and the remainder to vehicle impacts. This estimate takes no account of the 206 known fatalities of building occupants during the six years or of the unknown number of lesser casualties.

The only long-term trends that are clearly apparent and not simply related to changes in consumption of different fuels are linked to the conversion from manufactured (town) to natural gas. This has been associated, directly or indirectly, with a marked reduction in the risk of damage by an explosion resulting from a leak of gas within the building. It has however led to a small increase in the risk of damage by an explosion of gas entering a building from an external leak. An intensification of efforts to reduce the number of such leaks is therefore recommended.

Some attention is also given to seasonal variations in the risk of gas explosions, with particular reference to the spate of explosions in December 1976 and January 1977. This seasonal variation is shown to occur largely in the numbers of explosions of gas entering a building from outside with peaks closely correlated with a large preceding variation in soil moisture content and with unusually cold weather.

No truly comparable data have been found for other countries, but those that are available do not suggest that the British risks are unduly high.

Full details of all new BRE publications
are published quarterly in BRE NEWS.
Requests for BRE NEWS or for placing on
the mailing list should be addressed to:

Distribution Unit
Application Services Division
Building Research Establishment
Garston, Watford, WD2 7JR

ISBN 0 11 670755 0

First published 1978

## CONTENTS

# 1 INTRODUCTION

## 1.1 The purpose and scope of the survey

Damaging accidental explosions are fortunately not occurrences of which most people have direct experience – as they do, for instance, of strong winds. Partly for this reason, and partly because the damage to buildings rarely spreads far from the source of the explosion, they were not explicitly considered as a structural hazard in design until the Ronan Point collapse in 1968. After it they could no longer be ignored. Statistics then available at the Fire Research Station, and based on fire brigade reports, suggested that 1 in 50 blocks of flats of the size of Ronan Point might experience a similar explosion in the course of a 60-year life[1]. With the intention of ensuring that there should be no repetition of the extensive ensuing collapse, new legislation was introduced, to be followed by new provisions in Codes of Practice[2,3].

The available statistical data had not, however, been collected for the purpose of quantifying a structural hazard and there was a widespread feeling that fuller information was desirable. In October 1970 a pilot survey of accidental damage to buildings in south-east England was started by the Concrete Society and the Construction Industry Research and Information Association (CIRIA). This led to a contract being placed by the Department of the Environment with CIRIA for a two-year continuation of the survey from April 1971 to March 1973 with an extension of its geographic coverage to the whole of Great Britain. During this period the work was guided by an Explosions Working Party of the Construction and Housing Research Advisory Council's Structural Stability Panel[4], and was undertaken by a team drawn from the consulting engineers E W H Gifford & Partners and Ove Arup & Partners. On the conclusion of this contract, a further contract was placed with CIRIA by the Building Research Establishment (BRE). Initially this was for a further three years, but it has since been extended annually and is still in force. Under this contract, most of the collection of data has been undertaken by CIRIA itself in association with the Building Research Station. Continuity has been maintained both through the Building Research Station (which was similarly represented on the Explosions Working Party) and through the third of the present authors (who was a member of the previous team and has continued to act in a consultative capacity).

Attention has been concentrated throughout on accidental damage to non-industrial buildings. Damage due to deliberate bombings has been excluded; but damage due, for instance, to explosions resulting from attempts at suicide by gassing has been included since the explosions may be regarded as accidental. The restriction to non-industrial buildings has been interpreted in terms of the function of the building or building complex in which the explosion occurs. Thus damage to and resulting from explosions in certain service structures serving non-industrial sites has been included, whereas damage to non-industrial buildings on or adjacent to an industrial site as a result of a major industrial explosion (as at Flixborough) has been excluded. In the initial pilot survey, only damage due to explosions was considered, but it soon became apparent that another major cause of accidental damage was impact by vehicles. Accidental impact damage has therefore been included throughout the period now under review.

The prime purpose has been to quantify the structural hazard in terms of expected frequencies of different categories of load in different types of building as a possible guide to future legislation on structural design. In this respect the survey has differed from the other major survey of a similar kind undertaken over the same period – that of British Gas which was concerned more with the hazards associated with the use of a particular fuel[5]. Inevitably however, both have yielded data relevant also to the prime purpose of the other, and some attention will also be given below to the wider relevance of the present data.

## 1.2 The scope of this report

A report on the first two years of the full survey – covering the period April 1971 to March 1973 – was published in 1974[6]. It was originally intended that the present report should merely carry the story forward from April 1973. As a check on the maintenance of consistent standards of assessment, it was however considered necessary to review the data for all six years as a whole. This review, made in the light of a further four years' experience, showed the desirability of modifying some of the earlier assessments. The present report therefore covers all six years of the full survey up to March 1977 in the same detail and supersedes the earlier report, which is withdrawn.

As in the earlier report, tabular summaries are given of all reported non-industrial accidental explosions and vehicle impacts with non-industrial buildings throughout England, Scotland and Wales. Those incidents judged to have caused significant structural damage (as defined below) are then further analysed in relation to type, use, age and height of building and other factors that might have a bearing on the risk. Specific incidents and the lessons that can be drawn from them are not discussed. Numerous reports on major incidents have been prepared however, and one of these has been published[7], as has a summary of some others[8].

## 1.3 Sources of data

Press reports were the primary source of data. These were obtained through a press cutting agency and the incidents so identified are referred to below as *reported incidents*. The incidents were classified as the reports came in, the classification including a provisional assessment of 'significance' and 'severity' (in the senses discussed in the next section).

At the start of the survey the co-operation of the Gas Council (now British Gas) was obtained. Supplementary information has been obtained from this source on numerous incidents involving piped gas, and early notification has been received of some of the more severe incidents – a further source of such notification being telex messages from the Press Association.

The other principal source has been direct inspection of the resulting damage, in which connection early notification has been particularly valuable. These inspections were most numerous in the first year of the survey, when they covered more than half the incidents provisionally assessed as 'significant'. Subsequently they were made increasingly selectively up to the last quarter of 1976 and the first quarter of 1977, when they were again more numerous. In some cases, inspections were also made by the Midlands Research Station of British Gas and/or the British Ceramic Research Association and the Fire Research Station, and their findings also have been taken into account.

Final assessments of 'significance' and 'severity' have been made on the basis of all the above information. In addition we have had access to the independent statistical data of British Gas and the Fire Research Station. These, however, have been used only to check the reliability and adequacy of coverage of the press reports and to confirm doubtful details, and not to produce composite statistics.

### 1.4 Classification of incident and building types

All incidents were classified first in terms of the type of incident and the type of building (or buildings) affected.

Broadly there were three types of incident: those due to explosions of piped gas, those due to other explosions, and those due to accidental impact. Explosions of piped gas were broken down into explosions of town gas (ie manufactured), natural gas and liquefied petroleum gas (LPG). Other explosions were broken down into explosions of liquefied gas (gas cylinders and aerosols), high explosives (bombs), other materials (dust, chemicals, petrol spillages), and water containers (usually storage water heaters), with further categories of miscellaneous and unknown (the former including lightning). All impacts during the period covered were due to road vehicles, though they included one impact due to the overturning of a mobile tower crane that was not travelling at the time of the impact. Bomb explosions were included in the initial analysis of all reports received, but they were excluded from the subsequent analysis.

The classification in terms of building type was undertaken for the purpose of showing variations in risk with type and had to recognise both the desirability, for this purpose, of having enough incidents in each category to give a meaningful indication of risk, and the typical limitations of the press reports.

In the initial analysis of all reports received (Table 4), the limitations of the briefer reports on many minor incidents largely determined the classification that was adopted. The same classification was carried over into the first tabulation of 'significant' incidents (Table 6) to allow easy comparison, but for the further analysis of these incidents to give average risk figures (Table 8), some changes were made. Risks were calculated in terms of 'units' of accommodation, a unit being generally a single occupancy within a building but a single-type occupancy in certain cases, notably offices, where it was impossible to distinguish further. In these terms there were ample data to establish meaningful average risks for several different types of dwelling and this general category was therefore broken down into detached houses, semi-detached houses, terraced houses, bungalows, flats (including maisonettes) in purely domestic buildings, and flats over shops. Only the bungalow here was identical with one of the subdivisions of the general dwelling category in Table 4 – the single-storey dwelling. Conversely, some grouping of the earlier non-dwelling categories was necessary to establish meaningful risks. The groupings were made between categories of use likely to be associated with similar risks and involved some breakdown of the earlier 'other use (non-industrial)' category. In the initial analysis of all reports received, and only there, incidents in industrial buildings were also included.

### 1.5 Classification of the significance and severity of incidents

Categories of significance and severity were more difficult to establish and they became more clearly defined as the survey progressed. Since the main objective was to quantify a structural hazard not necessarily adequately provided for by normal design procedures and assumed loads, the categories should obviously be related as closely as possible to different higher levels of load. There were three major difficulties in so relating them. The first was the dynamic nature of all the loads. This meant that they could not be fully characterised by a single force or pressure parameter but only by a complete force–time or pressure–time relationship or, more generally, by a series of such relationships for different parts of the structure[8–10]. The second was the nature of the evidence available for assessing the actual loads – at best a fairly full picture of the damage caused to a structure of known characteristics before this evidence was disturbed by rescue operations and shoring or further demolition, but usually something well short of this ideal. Even in the ideal situation this evidence could usually suggest only upper or lower bounds to the likely loads. If the whole structure was of fairly low strength, its complete demolition might indicate only a misleadingly low lower bound, while a very strong structure would, conversely, show little evidence of a higher load that was still within the limits of this strength. The third difficulty arose not in the estimation and characterisation of the loads but from the direct dependences of their actual magnitudes and distributions on the strengths and certain other characteristics of the structures experiencing them. Impact loads depend in one way (for given impacts) on the resistances offered to the impacting masses; and the pressures developed in gaseous explosions depend in another way (for given quantities and distributions of gas and sources of ignition) on the impediments to free expansion of the burnt gas offered by the enclosures in which they occur[10]. These dependences are, of course, relevant to the use of the data obtained as a guide to the specification of loads to be withstood by future buildings with, perhaps, other characteristics.

In the face of these difficulties some compromise was necessary. A basic classification was made directly in terms of the type and extent of damage, recognising as far as possible the greater extent of damage to be expected from a given load on a weaker structure. Where, and only where, it was possible to inspect the damage and thereby gain adequate information, estimates were made of the peak pressures probably reached in gaseous explosions*; and, primarily to give some indication of overall costs to the country, estimates were made of the likely costs of structural reinstatement.

The classification that was finally adopted is illustrated in Figures 1–12 and was as follows:

A *moderate explosion* is one that destroys relatively weak structures up to and including the typical small bungalow or that, in the case of more substantial and larger structures, both destroys light cladding and windows and damages stud partitions and similar elements and blows doors off hinges.

A *moderate impact* is one that causes local collapse only (confined to a single storey and a similar width) of the brick walls of the typical post-1919 two-storey house or that does comparable damage to other types of building.

A *severe explosion* is one that destroys or seriously damages heavier claddings, infills, and partitions – for example of brickwork or blockwork – and similarly destroys or damages load-bearing elements such as the gable wall and first floor of the typical post-1919 two-storey house.

A *severe impact* is one that causes damage as in a moderate impact but of greater extent, or that causes local collapse only of stronger load-bearing elements – for example of reinforced concrete.

A *very severe explosion* is one that destroys or seriously damages stronger load-bearing elements – for example of reinforced concrete – or that causes very extensive damage over several storeys or to several buildings of less substantial construction.

A *very severe impact* would be one that caused more than local collapse of strong load-bearing elements or damage similar in extent to that caused by a very severe explosion to buildings of less substantial construction.

Different definitions are necessarily given for explosions and impacts because the gaseous explosion results in fairly uniform pressures over considerable areas, whereas an impact (like a bomb explosion) results in much higher pressures over small areas, so that the patterns of resultant damage are very different. One special class of explosion included in the survey is, however, more like an impact or bomb explosion in this respect. This is the explosion of a storage water heater due to excessive internal steam pressure. There are also similarities in the case of lightning, which can produce high local pressures within structural elements and thereby burst them open. The words 'would be' are used for the very severe impact because none was reported during the six years.

Finally, a *significant incident* is any explosion or impact falling in one of the above classes. Lesser incidents are referred to below only under the broader heading of *reported incidents*, which includes also the significant ones.

It will be noted that, according to the definitions, severe and very severe incidents could occur only in buildings of sufficient size and strength to provide the required evidence of severity – a consequence of the second difficulty referred to above.

The estimates of the peak pressures reached in some of the gaseous explosions were based to a considerable extent on observed breakages of window glass where this was clearly not due to impact by debris or frame distortion[11]. Implicit in this basis of estimation (apart from a considerable margin of error due to the wide variability of glass strengths) was the assumption of a peak of fairly short duration – shorter than the typical response period of major load-bearing elements of the structure. In general it may be assumed that equivalent static pressures for such elements would be from 20 to 40 per cent less, and where failures of such elements have been used as a basis of estimation, the direct estimates have been suitably enhanced for consistency. It would be reasonable to assume that the complete effective pressure pulses were, as far as could be ascertained, dynamically equivalent to pulses of displaced-cosine profile, of a peak value within the range given, and of a duration of about 0.25 s. The available experimental evidence suggests that actual pressure pulses will usually depart considerably from this rather arbitrary norm[12–15]. The estimates relate to the pressures generated near the source of the explosion. The one exception is the second estimate (shown in parentheses in Appendix A) for incident number 1996. This incident was fully inspected and showed clear evidence of a second shorter pressure pulse of higher peak value in another part of the building[7]. It is to that that the second estimate relates.

The estimates of costs of structural reinstatement took account, as far as possible, not only of the extent of damage (including damage to neighbouring buildings) but also of the ease or difficulty of the necessary works. They were also discounted, to some extent, for betterment in the case of old buildings not constructed to current standards. They were all made in terms of current costs at the time of the incident. Since, over the six years, there was an average rise in costs of about 150 per cent, the costs for later years are considerably inflated in relation to those for earlier years.

*Estimates previously made for other incidents have been deleted.

**Figure 1** Destruction of the outer brick skin of the front of the house, windows and doors blown out, and some internal damage (Reproduced by courtesy of the Echo, Sunderland)

**Figure 2** Destruction of a length of external wall on one floor plus some internal damage to ceilings, etc (Reproduced by courtesy of Southern Newspapers Ltd, Southampton)

## Moderate impacts

**Figure 3** Hole punched in the front of a two-storey house by an articulated lorry after a collision (Picture by courtesy of Hull Daily Mail)

**Figure 4** Partial destruction of a bungalow by a lorry (Reproduced by courtesy of Eastern Daily Press, Norwich)

**Figure 5** Very extensive damage to four flats in a two-storey terrace (Reproduced by courtesy of Hall Photography, Greenock)

**Figure 6** Collapse of the gable wall and partial collapse of the roof of a two-storey semi-detached house (Reproduced by courtesy of Greater London Council)

## Severe explosions

**Figure 7** Partial collapse extending over three storeys of a modern three-storey block of flats (verging on very severe) (Reproduced by courtesy of Nottingham Evening Post)

## Severe impacts

**Figure 8** Partial collapse of the gable wall and first floor of a two-storey shop plus flat as a result of a lorry impact (Reproduced by courtesy of the Telegraph and Argus, Bradford)

**Figure 9** Collapse of almost the entire front and part of one gable and partial collapse of the first floor of a two-storey house (Reproduced by courtesy of Southern Newspapers Ltd, Southampton)

## Very severe explosions

**Figure 10** Total destruction of two buildings of five and six storeys plus further extensive damage to adjacent buildings. These were late-nineteenth century buildings with timber floors throughout (Reproduced by courtesy of South West Gas)

**Figure 11** Complete destruction of heavily reinforced concrete walls and severe damage to reinforced concrete beams in a void beneath a terrace of shops

**Figure 12** Severe damage to reinforced concrete slabs and walls at the foot of a 15-storey block of flats

**1.6 Checks on the validity of the data**
'One cannot believe all that one reads in the papers.' This is certainly true of many reports of buildings or parts of buildings being 'destroyed'. But once a tendency towards somewhat exaggerated reporting has been recognised, it need no longer mislead. Assessments, where not made on the basis of actual inspection of the damage, were made only in the light of inspections of other incidents and on the basis of such factual evidence (particularly photographs) as was included in the reports. Thus any errors of wrong assessment of the severity of reported incidents are almost certainly underestimates due to insufficient evidence rather than overestimates.

A further possible source of error that gave more cause for concern was the length of the reporting chain. Incidents had first to be reported to a newspaper; the editor had then to accept the report for printing; and the cutting agency had finally to spot and forward the report. It was certainly possible that the reports finally received would not cover all significant incidents, thus leading to further underestimation of average risks.

To check on these possibilities, comparisons were made between the yearly totals of explosions in different categories as obtained from the present survey and comparable totals obtained independently from reports made directly to the Home Office on all fires attended by fire brigades (which should mention explosions where they are believed to have occurred) and to British Gas on all incidents involving piped gas. For one year a more detailed comparison was made with the fire brigade reports to identify more precisely some shortcomings of the present data. In addition several internal checks were made on the consistency of assessments of significance and severity.

The initial comparison with analyses made at the Fire Research Station of the fire brigade reports is reproduced in Table 1. These figures relate only to incidents involving piped gas in dwellings and fully overlap, unfortunately, by only three years because the Fire Research Station analyses are available only about two years in arrears. There is no exact correspondence between the respective categories of 'significant' incident. For the three years 1972–74, the most nearly comparable category that was available in the standard coding of the reports after they are received from the Home Office was that of an explosion that involved damage beyond the room of origin. It was therefore the totals of these incidents that were taken. The fact that they are somewhat higher than the corresponding survey figures is to be expected and it is possibly most significant that they show comparable fluctuations from year to year. The usually greater numbers of reported incidents could well indicate either more reports of very minor incidents or a proportion of reports in which a minor explosion accompanied or led to a fire in the normal sense which was reported in the press (if at all) only as a fire.

In view of the inconclusiveness of this comparison, a further more detailed comparison was made, by courtesy of the Fire Research Station and the Home Office, for 1974 only. For this purpose individual reports on 'significant' incidents were reviewed. The result of this review was to disclose gaps in both sets of reports. Three explosions (two of piped natural gas and one of liquefied gas, and all in dwellings) on which no press reports had been received were identified as probably having been of 'moderate' significance. Eight explosions (including two of piped natural gas and three of liquefied gas and all but one in or also involving dwellings) on which press reports had been received, but with insufficient detail to justify a 'significant' rating, were also identified as probably having been of 'moderate' significance. A further two explosions (both of piped natural gas in dwellings) on which similarly inadequate press reports had been received were considered to have probably been in the 'severe' category. The addition of all these incidents would increase the total of 'significant' explosions for the year from the 51 identified from the press cuttings to 64 and, taking into account the fact that the Fire Research Station/Home Office data were also found to be incomplete, the true total might even be slightly higher.

The comparison with the data collected by British Gas on explosions of piped gas is made in Table 2 on the basis of the figures given in Tables 1 and 2 of the King Report[5]. The British Gas figures for numbers of explosions causing fatalities and numbers of fatalities should be virtually identical to those from the present survey since differences are to be expected only in a few borderline cases of delayed fatalities and fatalities associated with, but not necessarily caused by, explosions. If allowance is made for these differences the agreement is good – as might well be expected since fatalities are unlikely to escape the notice of either the press or a public utility that may be involved.

Agreement is less close between the respective figures for all 'significant' explosions and those that are 'severe' or worse. In the first case, there is no doubt that the considerably higher British Gas figures result largely from a different and broader criterion of 'significance'. This was purely in terms of the cost of the damage caused and, with a lower limit of £100 cost, was well below the lowest estimates of costs of structural reinstatement for the explosions classed as 'significant' in the present survey. The British Gas 'severe' category was more directly comparable. It meant partial or total demolition of premises, and was therefore only a little broader in including some cases of partial demolition, and some of total demolition of relatively weak structures, that were rated as only 'moderate' in the present survey. Bearing the differences in definition in mind, the most that can be concluded from these latter comparisons is probably that they are in no way inconsistent with the broad indications of a certain incompleteness of coverage in the present survey given by the detailed comparison with some of the Fire Research Station/Home Office data.

The internal tests on consistency of assessment were of two kinds. First, the estimates of costs of reinstatement, suitably adjusted for inflation, were compared with the severity ratings. Considerable overlap was to be expected between the ranges of cost estimates for the different categories for several reasons touched on above. Some estimates nevertheless seemed unduly high or low in relation to the initial rating, and in these cases the evidence was carefully re-examined. They usually proved to be borderline cases and some re-assessments were made,

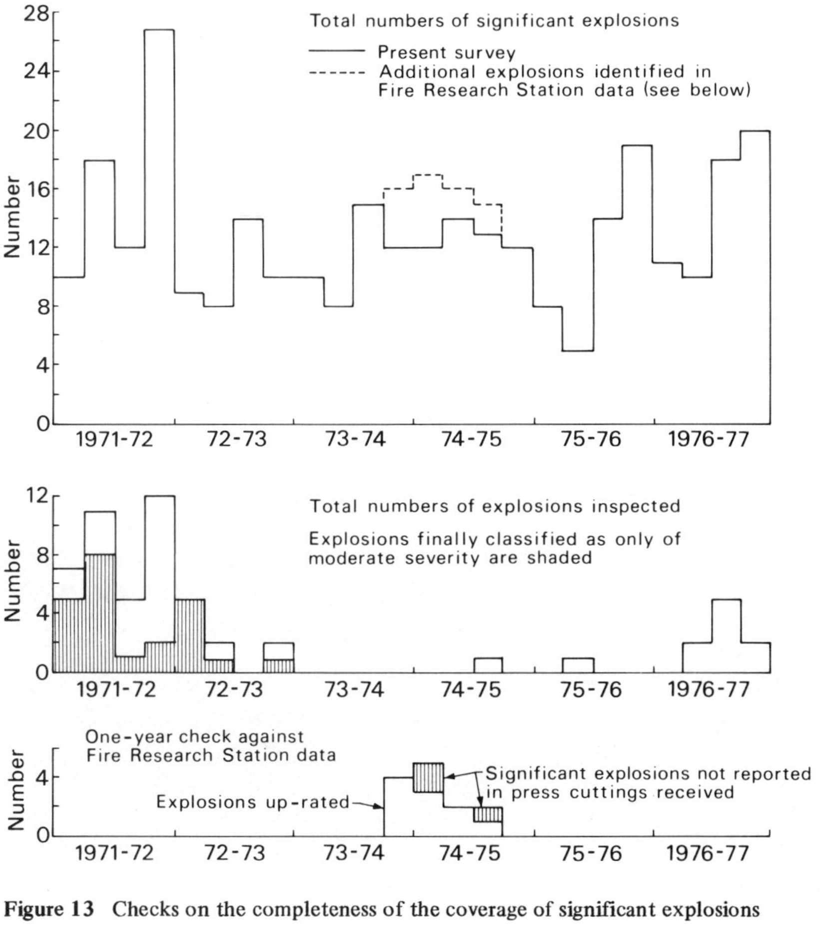


**Figure 13** Checks on the completeness of the coverage of significant explosions

chiefly by down-grading some of the earlier incidents originally assessed as 'moderate'. Secondly, and in parallel with the detailed comparison with Fire Research Station/Home Office data for 1974, the possibility that the widely varying proportion of the total number of reported incidents that was inspected had influenced the numbers assessed as 'significant' was considered. This was certainly found to have been the case in one way – several of the borderline incidents just referred to as having been down-graded proved to be ones that had been inspected and that would not finally have been assessed as significant on the basis of the press cuttings alone. The making of an inspection seems to have led to a slight, and understandable, bias towards retention of the incident in the main statistics. A few other cases were found where an inspection had indisputably shown that an incident should be placed in one of the 'significant' categories in spite of insufficient evidence in the press reports to justify this.

Figure 13 summarises the most relevant evidence bearing on the completeness of coverage of the data presented in the remainder of this report. It suggests that numbers of significant explosions may have been underestimated by some 20 per cent in the middle years of the survey and by some lesser percentage in the first year and towards the end, primarily as a result of inadequate coverage by the press cuttings received. It must also be relevant, in this connection, that public interest, and hence news value, have varied considerably. The interest generated by the Ronan Point incident probably carried over into 1971, and there was a marked renewal of interest late in 1976. Against the inadequacies of the press as a source of information that it does not really set out to provide, must be set, however, the valuable immediacy of its coverage and its extent to all kinds of incidents.

No similar checks on the extent to which accidental impacts were covered were found to be possible. Numerous bodies were approached in the hope of finding comparable statistics collected from an independent source, but no other statistics were found.

The much larger variations in the yearly totals, both of reports received and of 'significant' incidents, strongly suggest even larger changes in press interest than in the case of explosions. Other indications of incomplete coverage were the receipt of reports that referred also to previous impact damage to the same building when no report of the previous impact could be found on looking back through the records, and the non-receipt of reports from the agency on incidents of which we had other knowledge. It was concluded already in the report on the first two years that impacts with buildings were unlikely to be reported unless the results were spectacular or people were killed or injured. This would, indeed, be consistent with the apparent practice in reporting impacts between vehicles.

Rather than, for these reasons, omitting any full analysis of the significant reported impacts as in the previous report, we have however included some such analysis here with cautionary notes about the likely shortcomings of the data. It can be said with confidence only that the risks are no less than those indicated.

## 2 FREQUENCIES AND ESTIMATED COSTS OF INCIDENTS

### 2.1 Reported incidents

Tables 3a–c give quarterly, annual and cumulative totals of all reported incidents classified according to type of incident. It will be seen that about a quarter of the total number of reports were of explosions of piped gas. Of the other single categories of incident, explosions of liquefied gas (gas cylinders and aerosols) accounted for the next largest number of reports, and after them vehicle impacts. On average about 5 explosions of piped gas, 3 of liquefied gas and 2.5 vehicle impacts were reported per week. There were some considerable fluctuations in the numbers of reports from year to year, particularly of vehicle impacts, but it is not considered that much significance should be read into these fluctuations: they may well be largely due to fluctuations in what editors judged to be the threshold of news interest.

In Tables 4a–d the total numbers of reports received over successive two-year periods and over the full six years are further analysed in terms of the use of the building damaged in the incident or, where there was more than one use, in terms of the use primarily associated with the explosion or affected by the impact. The heading 'external' is included to cover those reports that were received of, for instance, explosions on rubbish dumps, external explosions of gas leaking from mains, and vehicle impacts with boundary walls or fences without, in each case, any mention of damage to buildings.

As would be expected from the numerical preponderance of two-storey dwellings in the total building stock, most incidents of all kinds except the unknown were reported as occurring there. If the 'external' incidents are ignored as not really relevant, shops, cafes and public houses together were the subject of the next largest number of reports.

### 2.2 Significant incidents

From the fuller information usually recorded about each significant incident, details of the use, type, age and structure of the building, and the main details of the nature, severity and cost of the incident have been tabulated individually in the Appendices. Incidents are identified there only by numbers (in two sequences), but the full records also give date and place, and time of day where known.

Tables 5a–c give quarterly, annual and cumulative totals in the same manner as they were given in Tables 3a–c for all reported incidents. Tables 6a–d likewise give comparable figures to those in Tables 4a–d. In addition the annual totals given in Tables 5a–c are broken down into separate totals of moderate, severe and very severe incidents in Table 7.

Tables 5a–c show that, if attention is confined to significant incidents, piped-gas explosions accounted for almost half the total number reported, with vehicle impacts a close second and explosions of liquefied gas now far behind in third place. Table 7 shows that, if it is further confined to severe and very severe incidents, piped-gas explosions accounted for three-quarters of the total number of severe incidents and for all those judged to have been very severe. The importance of vehicle impacts is then, apparently, greatly diminished, since they accounted for only a little under a tenth of all severe incidents. But the fact that nearly half the severe impacts were reported in one quarter (the last quarter of 1972 – Appendix B) throws further doubt on the reliability and adequacy of the press reports for assessing this risk and strongly suggests that fuller reporting would have disclosed more such incidents.

These tables also show that, on average, 1.64 significant incidents were reported per week, made up of 0.76 explosions known to be of piped gas (including 0.30 that were classified as severe or very severe), 0.105 explosions known to be of liquefied (cylinder) gas (including 0.02 classified as severe), 0.095 accidental explosions of other known kinds (including 0.04 classified as severe), 0.03 accidental explosions of unknown kinds, and 0.65 impacts (including 0.03 severe ones).

Finally, a comparison between Tables 4 and 6 shows that an even larger proportion of significant incidents (as compared with that of all reported incidents) occurred in two-storey dwellings. The next highest proportion again occurred in shops, cafes and public houses, but was roughly half that for all reported incidents. It is possible that the latter difference reflects a greater press interest in damage to buildings of these kinds and hence a fuller coverage of relatively minor incidents than in the case of incidents occurring in dwellings.

### 2.3 Estimated costs

The estimated costs of structural reinstatement ranged widely and, even after checking, there remained considerable overlap between the ranges for incidents of different assessed severities. If the estimates are adjusted to price levels at the end of the period of the survey[16], the ranges for moderate, severe and very severe explosions were approximately £800 – £20 000, £2000 – £60 000, and £20 000 – £300 000 respectively. For moderate and severe impacts (on the basis of smaller samples) they were approximately £400 – £5000 and £4000 – £10 000 respectively. The lower estimates for impacts are to be expected since impact damage of a given severity tends to be more localised. There may however have been some underestimation here on account of the difficulties of identifying consequential damage away from the impact itself from the evidence available.

Annual estimated costs, again adjusted to final price levels, varied from approximately £350 000 to £950 000 for explosions and from approximately £10 000 to £100 000 for impacts. The average costs were approximately £550 000 for explosions and £50 000 for impacts.

In considering these costs in relation to likely costs of possible preventive measures, the following facts should be borne in mind:

1 There is evidence, as noted in Section 1.6, that some significant incidents have been missed. If all had been included, the average annual costs might have been nearer £600 000 for explosions and, say, £80 000 or more for impacts.

2 Nothing has been included for 'non-significant' incidents. There were, on average, 780 such explosions reported per year. If the average cost of structural reinstatement for these is taken as £100 at final price levels, almost another £100 000 must be added to the average annual cost. The addition for non-significant impacts might bring the average annual cost here up to, say, £100 000.

3 The estimated costs are for structural reinstatement only. The total costs, excluding those of personal injuries and deaths, might well have been double these costs.

When these facts are taken into account, the best estimate of the total annual cost to the country, *excluding the costs of personal injuries and deaths*, is about £1½ million, with between 80 and 90 per cent of this attributable to accidental explosions and the remainder to vehicle impacts. The figure would have been substantially higher if the cost of the damage to non-industrial buildings that resulted from the Flixborough explosion and a few other industrial explosions had been included.

Costing of the personal injuries and deaths is more difficult. If a value of £100 000 is placed on each life lost, the annual cost of the deaths alone was more than double that of the material damage, since there were 206 associated fatalities over the six years excluding vehicle drivers and passengers. This suggests an inclusive annual cost to the country of over £5 million.

## 3 FURTHER ANALYSIS OF SIGNIFICANT ACCIDENTAL EXPLOSIONS

### 3.1 Risks in terms of explosion pressures

Figure 14 gives some indication of the frequencies with which pressures of different magnitudes are likely to be reached. Since it was only during the first two years that a fairly large and representative selection of significant incidents was inspected, the two curves relate only to the respective maximum and minimum estimates of peak pressure (as defined above) for these incidents. There were 44 inspections of significant explosions in all, as compared with a yearly average for the six years of 51.5, so the numbers of incidents have been extrapolated, on a second scale, to this slightly larger total. If anything,

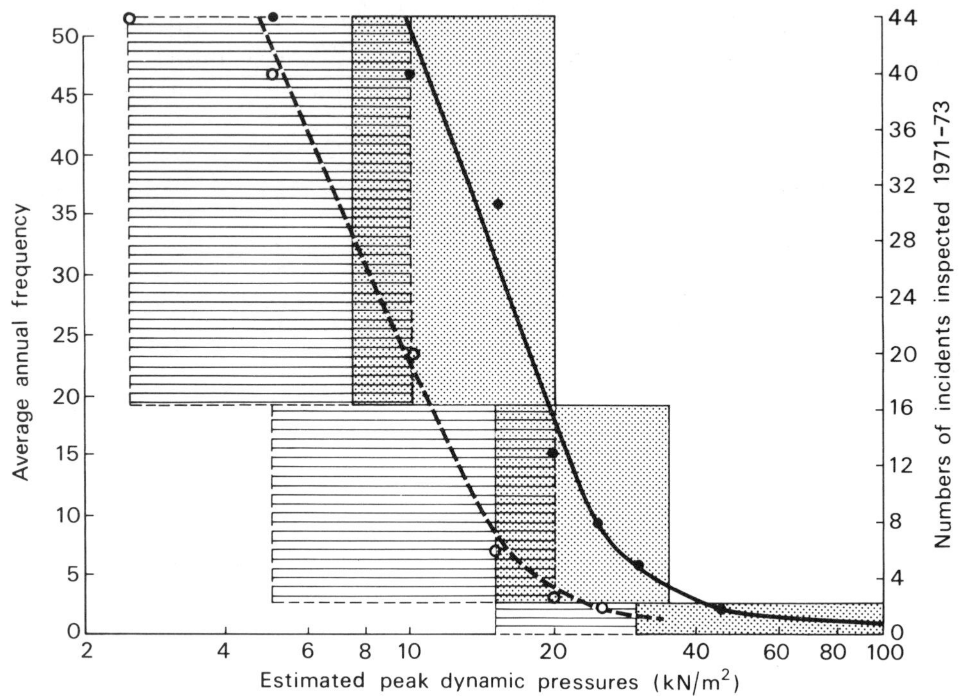


**Figure 14** Frequencies with which given peak dynamic pressures may be reached or exceeded in significant accidental explosions. Points relate to incidents that were inspected during the first two years, as do the accompanying curves. Full lines denote maximum estimates of the peak pressures; broken lines denote minimum estimates of the peak pressures

the risk of explosions in which the higher pressures are reached is probably slightly overestimated because the 44 incidents were made up of 4 very severe, 16 severe and 24 moderate as compared with an average make-up over the six years of 5 per cent (2.28) very severe, 32 per cent (14.10) severe and 63 per cent (27.62) moderate.

This probable slight overestimation is directly indicated by the superimposed blocks that relate to all moderate, all severe, and all very severe explosions. The blocks are bounded vertically by the total number of explosions in each category and horizontally by the extremes of all maximum and all minimum estimates of the peak pressures throughout the six years. There were, for instance, 194 moderate explosions and, for those that were inspected, the maximum estimates of peak pressure ranged from 7.5 to 20 $kN/m^2$ and the minimum estimates from 2.5 to 10 $kN/m^2$. Each curve should lie wholly within the corresponding three blocks. (It will be noted, in this connection, that the highest points corresponding to the lowest estimated pressures were ignored in drawing the curves. This is because of the overlaps of pressure ranges for different categories of severity which must mean that a considerable number of 'non-significant' explosions would also have reached these lowest pressures. Their inclusion would move the points considerably higher.)

In relation to the reference in the Building Regulations[2] to a pressure of 34 $kN/m^2$, it is of interest to estimate the frequency with which an equivalent static pressure of this magnitude is likely to be reached in an accidental explosion. Assuming that an average equivalent static pressure is 25 per cent less than the estimated dynamic peak pressures and taking the mean of the maximum and minimum estimates of the dynamic peak, the frequency is about once per year on the basis of the curves. During the six years there were two incidents in which this pressure was definitely exceeded, there was one borderline case (where the appropriate reduction factor was probably considerably higher), and there were nine very severe incidents that were not inspected as compared with seven that were inspected. The likely frequency of accidental explosions giving equivalent static pressures of half the specified magnitude is about five per year.

It is also of some interest to compare the estimates of pressures reached in explosions of natural gas with those for explosions of town gas. Of the 44 incidents that were inspected during the first two years, 17 were explosions of natural gas and 23 of town gas. The latter included, however, the very severe explosion at Clarkston Toll for which there was no counterpart in the former. If this incident is excluded and the mean estimates of peak pressure for all other incidents of each kind are averaged, the averages are approximately 11 $kN/m^2$ for the natural-gas explosions and 13 $kN/m^2$ for the town-gas explosions. While it is not considered that much weight should be placed on the small difference between these averages, there is clearly no basis here for attributing any higher risk to the use of natural gas and the conclusion of the Morton Report 'that the explosion . . . hazard associated with natural gas is similar to the hazard associated with town gas'[17] is, to this extent, confirmed.

**3.2 Risks in relation to type and use of building**

The risks of significant structural damage clearly vary considerably with type and use of building. In Tables 8a–d the average annual risks over successive two-year periods and over the whole six years are calculated in terms of both numbers of significant explosions and numbers of units of accommodation affected. In both terms, the risk averaged over the whole six years varied from 1.5 per million units per year for detached houses to a much more approximate figure of 16.7 per million units per year for the non-dwelling group consisting chiefly of public utility sub-stations and telephone exchanges. For dwellings as a whole the two risk figures, similarly averaged, were 2.1 and 3.2 per million units per year respectively; and for all non-dwelling categories as a whole they were 6.8 and 10.4 per million units per year respectively. The two risk figures differed most for flats over shops on account of the frequency with which significant damage was done by explosions that occurred or originated in the shop. The second figure here (for risk of damage irrespective of the location of the explosion) is comparable with that for damage to shops, etc. Perhaps the most notable feature is the extent of the variation of the second figure within the general dwelling category – from 1.5 to 4.3 per million units per year averaged over the six years even excluding the higher figure for flats over shops. A further factor to be borne in mind in relation to these tables as a whole is, however, the possibly considerable variations between the average sizes of 'units' of accommodation in the different categories.

### 3.3 Risks in relation to age of building

Table 9 shows (for dwellings only) a clear correlation between risk and age of building. Dwellings built before 1919 suffered about 40 per cent more significant explosions in proportion to their number than dwellings built more recently. Since allowance was made in assessing significance for the lower strengths of some pre-1919 buildings, the increased incidence cannot be attributed to this cause. Indeed if the criterion were merely extent of damage, the increase in risk might be substantially more than 40 per cent. Possible reasons for the 40 per cent increase are different patterns of occupancy, more cellars and other voids where gas could accumulate undetected in the older buildings, differences in the average age and condition of gas installations, and closer proximity to street mains.

### 3.4 Risks in relation to height of building

Table 10 (again for dwellings only) shows a further correlation between risk and height of building. In this case it is less firmly established because of the relatively small numbers of dwellings in buildings of over three storeys – only about 5 per cent of the total stock in all – and the lack of a more precise value of this percentage. But the fact that 6½ per cent (more precisely 6.3 per cent) of significant explosions occurred in this 5 per cent of the stock must indicate a substantially higher risk, even if the true figure is not 26 per cent greater as is suggested. An increase in risk of around 15 per cent is to be expected, on the basis of Table 8d, from the fact that all dwellings in buildings of over three storeys are likely to be flats or maisonettes (which, for the present purpose, have been considered as flats). But any further increase must stem from factors more directly associated with height. These might again include different patterns of occupancy. Another possible factor is the greater amount of internal ducting and internal service pipework in tall buildings.

### 3.5 Risks of gaseous explosions in relation to consumption and nature and source of gas

It will be seen from Tables 6d and 7 that at least 87 per cent of all significant explosions were of piped or bottled gas and that all very severe explosions were of piped gas. These gaseous explosions are analysed further in Tables 11 and 12.

If important factors like the ventilation provided (which might be expected to be broadly similar whatever gas is used) are set aside, the number of explosions of any particular gas might be expected to vary primarily with such other variables as the number of appliances in use, the frequency with which they are used, and the total length and characteristics of the distribution system. But, from the user's point of view, the most important relevant variables are probably the related ones of total consumption (in terms of energy equivalent) and number of consumers. The annual totals of all significant explosions of town gas, natural gas and bottled gas are expressed, as far as possible, in these terms in Table 11.

The most striking conclusion to be drawn from Table 11 is that users of bottled gas run a much higher risk, in proportion to energy consumed, than do users of piped gas. The risk is certainly less precisely established on account of the smaller number of explosions and the less accurate figures for consumption, but there can be no doubt that it is a whole order of magnitude greater. Piped gas is shown to be a relatively safe fuel, and it would clearly be entirely wrong to ban its use in tall buildings for instance, with the objective of making accidental explosions less likely, if there were any risk of some occupants turning to bottled gas as a substitute.

It also appears that natural gas is somewhat less hazardous than town gas, though the difference in risk here is much smaller and is possibly a little exaggerated by the fuller coverage (referred to in Section 1.6) in the first year of the survey when half the significant town gas explosions were reported. It is possible that one reason for the lesser risk associated with natural gas is the renewal of many appliances and the testing of installations that accompanied the conversion from town gas. It may also be relevant that a significant part of the overall increase in consumption (which is, of course, now largely of natural gas) has been due to increased use in central heating installations which are probably safer, in terms of energy consumed, than most intermittently used appliances. Nevertheless the figures again support the conclusion of the Morton Report that was referred to in Section 3.1.

Apparent fluctuations in the risks from year to year must be due, to some extent, to the variable coverage of significant explosions. To compensate for this the risks from 1972 onwards should be slightly increased. If they are increased as suggested in Section 1.6, the overall risk for all piped-gas explosions has remained reasonably constant since 1972 (when conversion to natural gas was around 50 per cent) at about 0.06 explosions per $10^9$ MJ.

Figure 15 shows corresponding data for dwellings only, including that obtained from the Fire Research Station for earlier years. The Fire Research Station/Home Office data suggest a risk fluctuating around about 0.1 significant explosions per $10^9$ MJ – the same as the average risk found in the present survey for all non-industrial explosions of town gas. It is not directly comparable though, on account of the different criterion of 'significance'. Making the necessary adjustments on the basis of the comparisons in Table 1 and in accordance with the suggestions of Section 1.6, the overall picture seems to be that significant explosions as here defined have occurred in dwellings over the past 20 years at rates of around 0.08 per $10^9$ MJ of town gas consumed and around 0.06 per $10^9$ MJ of natural gas. The apparent levelling-off of risk in relation to consumption is closely linked with the conversion to natural gas, but the more fundamental causes may well be the associated changes in gas installations discussed above.

Table 12 brings to light the one disturbing aspect of the overall reduction in risk associated with the conversion to natural gas. It shows that, of all piped-gas explosions for which the source of the leak could be established, 42 per cent resulted from leaks external to the building or at least on the service side of the meter, and that the overall reduction in risk with natural gas did not stem from any reduction in incidents of this kind. This is disturbing because they are among the types of incident that building owners and occupants can do least to prevent.

Forty-eight per cent of natural-gas explosions as compared with only 25 per cent of town-gas explosions made up the overall ratio of 42 per cent just referred to, these figures corresponding to an unchanged absolute risk of 0.021 significant explosions per $10^9$ MJ as a result of external leaks. If attention is concentrated on the severe and very severe explosions, even the absolute risk is seen to have worsened. In percentage terms it becomes 57 per cent for severe or worse and 73 per cent for very severe natural-gas explosions, while remaining at 25 per cent for both groups of town-gas explosions. In absolute terms it becomes 0.011 and 0.0023 per $10^9$ MJ for the two groups of natural-gas explosions as compared with 0.010 and 0.0017 per $10^9$ MJ for the two groups of town-gas explosions. A corollary of this slight worsening of the absolute risk of explosions due to external leaks is, of course, that the reduction in risk of explosions due to internal leaks has been somewhat better than that indicated by the overall risk figures quoted above.

Possible reasons for some increase in leaks from mains and service pipework are the higher supply pressures for natural gas and the natural dryness of the gas which can dry out joints in pipework, though it is understood from British Gas that there is little supporting evidence for any increase as a result of the higher supply pressures. Discussion with British Gas in 1976 also showed that, of 18 severe and very severe explosions of piped gas that resulted from external leaks during 1973–76, 7 could be attributed to breaks caused by interference by third parties. This usually took the form of excavations or building works in the vicinity of a main or service pipe. These topics are discussed further in the King Report[5] (where the risks are assessed a little differently on the basis of different ranges of severity of incident).

Figure 16 presents a composite picture of the make-up of the quarterly totals of significant explosions in each of the four main groups of building types considered in Tables 8a–d. It draws attention again to the high proportion of natural-gas explosions that resulted from external leaks and shows that it included nearly all such explosions in the shops group. Marked variations are also apparent in the proportions of explosions of different types in the different groups of buildings. These variations mean that the figures for average risks in Tables 8a–d and 11 are not independent of one another.

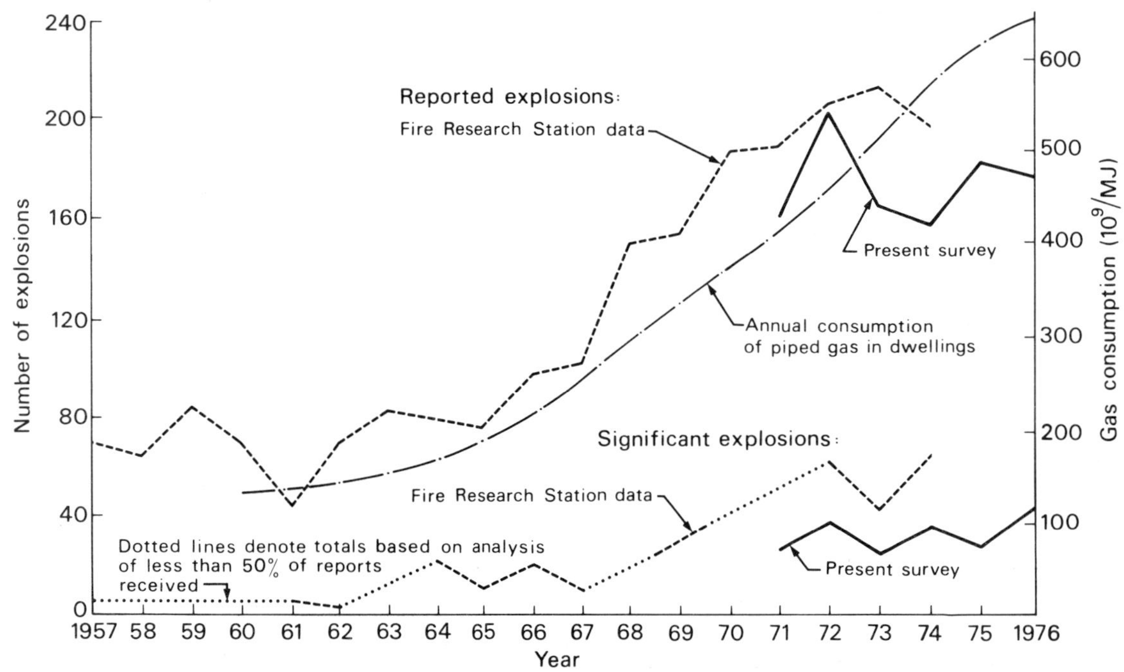


**Figure 15** Calendar-year totals of reported and significant piped-gas explosions in dwellings from Fire Research Station/Home Office data and from the present survey in relation to gas consumption

### 3.6 Seasonal variations in risk

Individual explosions, even of piped gas, do not however result merely from undetected or unattended leaks. The leaking gas must (if the leak is external) enter a building; it must in any case accumulate in sufficient quantity in one or more rooms or internal spaces; and it must be ignited while the gas–air mixture is of a flammable concentration at the point of ignition. Explosions of other kinds are also, in general, the result of unfavourable combinations of circumstances. It is not therefore to be expected that explosions of a given kind will occur at regular intervals determined only, for instance, by the rate of consumption of a given fuel. Considerable short-term random fluctuations in intervals and frequency are, on the contrary, to be expected. Such variations may certainly be seen in Figure 16.

The one aspect of the fluctuations in frequency apparent from this figure that cannot be considered just random, is the consistently greater frequency of explosions of piped gas from an external leak during the last two quarters of a survey year – the winter quarters extending from October to March. To investigate this seasonal variation further, and to check on any possible

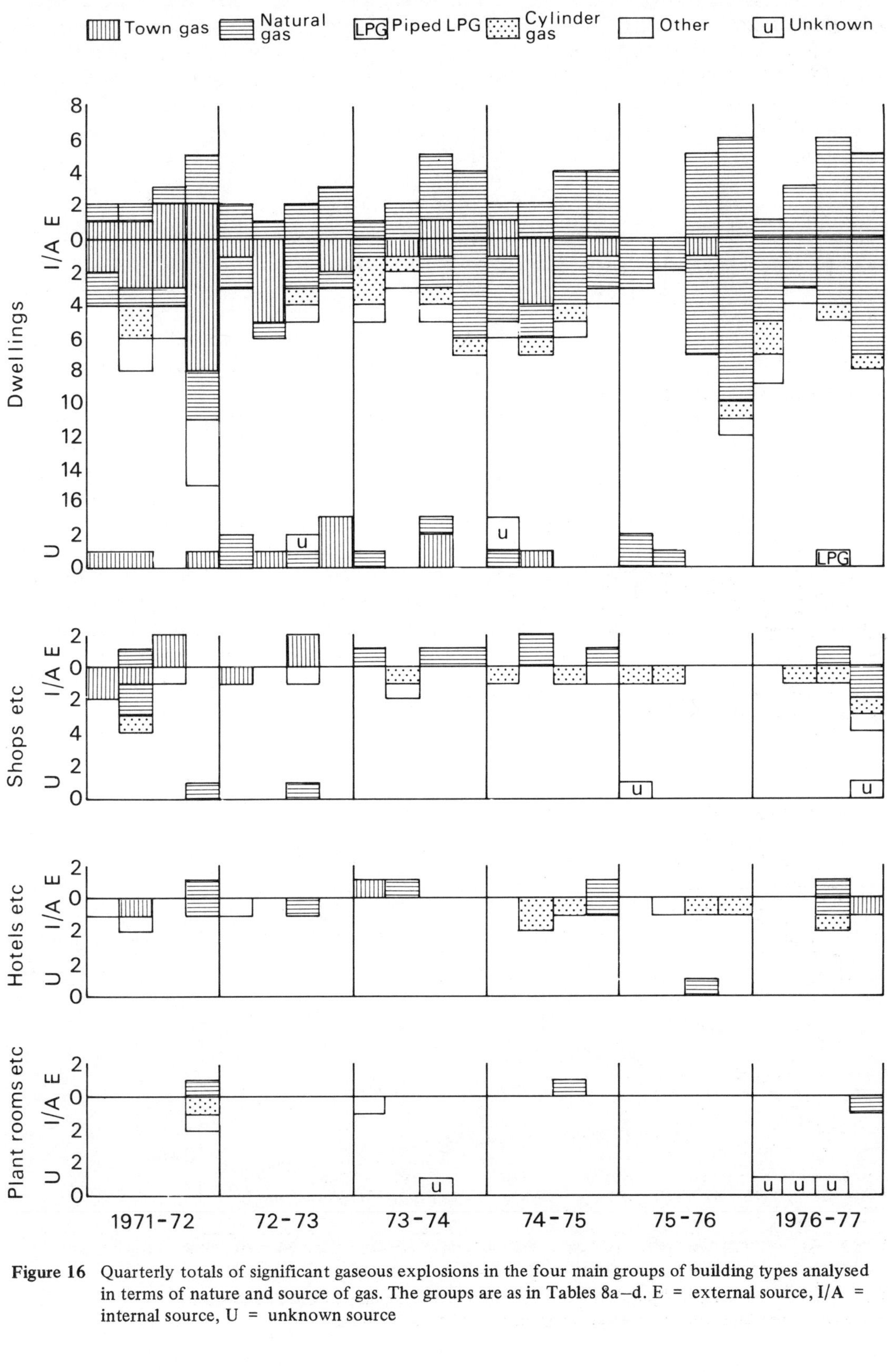


**Figure 16** Quarterly totals of significant gaseous explosions in the four main groups of building types analysed in terms of nature and source of gas. The groups are as in Tables 8a–d. E = external source, I/A = internal source, U = unknown source

seasonal variation in frequency of explosions of piped gas from an internal leak, the relevant data have been re-plotted in Figure 17 on a monthly basis. The numbers of significant explosions of each type have been aggregated over the six years for each month of the year. In addition the numbers for the final year (1976–77) have been separately plotted and compared with the six-year averages. The winter peak in explosions resulting from external leaks is seen to be more pronounced even than Figure 16 suggested. A much smaller seasonal variation is disclosed in the frequency of explosions resulting from internal leaks.

Among the seasonally varying factors that might be expected to be relevant in both cases is a common tendency to reduce ventilation in the colder winter weather. This will allow escaping gas to accumulate more rapidly. The importance of this factor is perhaps borne out by the more marked fluctuations in 1976–77 when there was an unusually large difference between winter and summer temperatures as shown in the top plots of Figure 18. The net effect, however, is less than might be expected which suggests that other factors may also be operative.

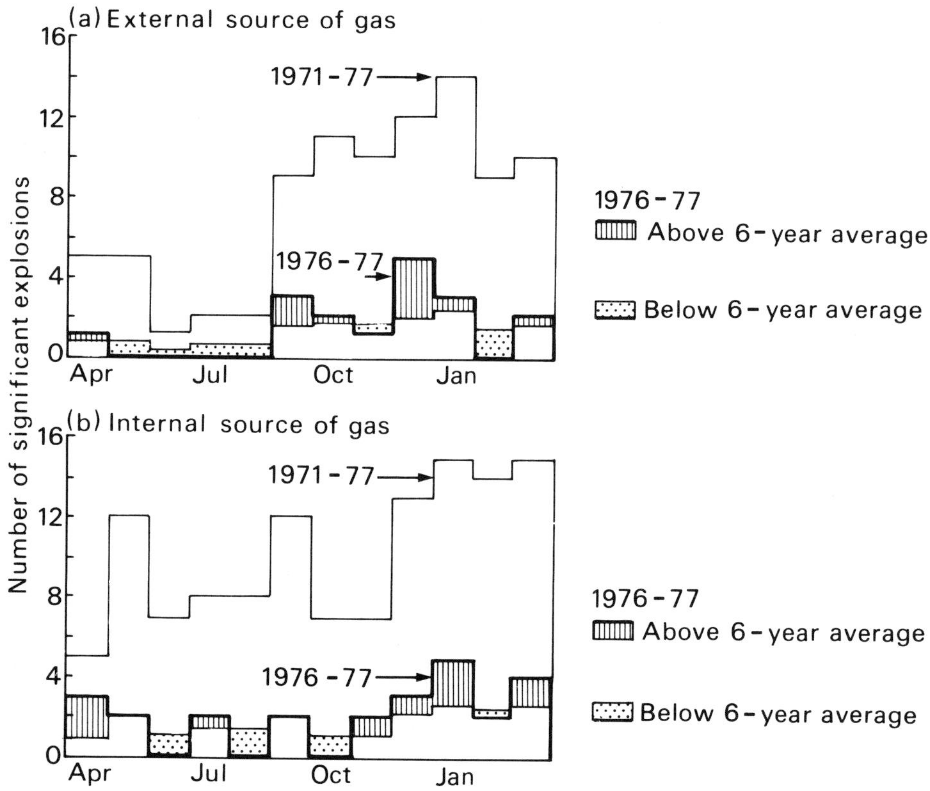


**Figure 17** Monthly totals of significant piped-gas explosions resulting from external and internal leaks respectively: 1971–77 and 1976–77

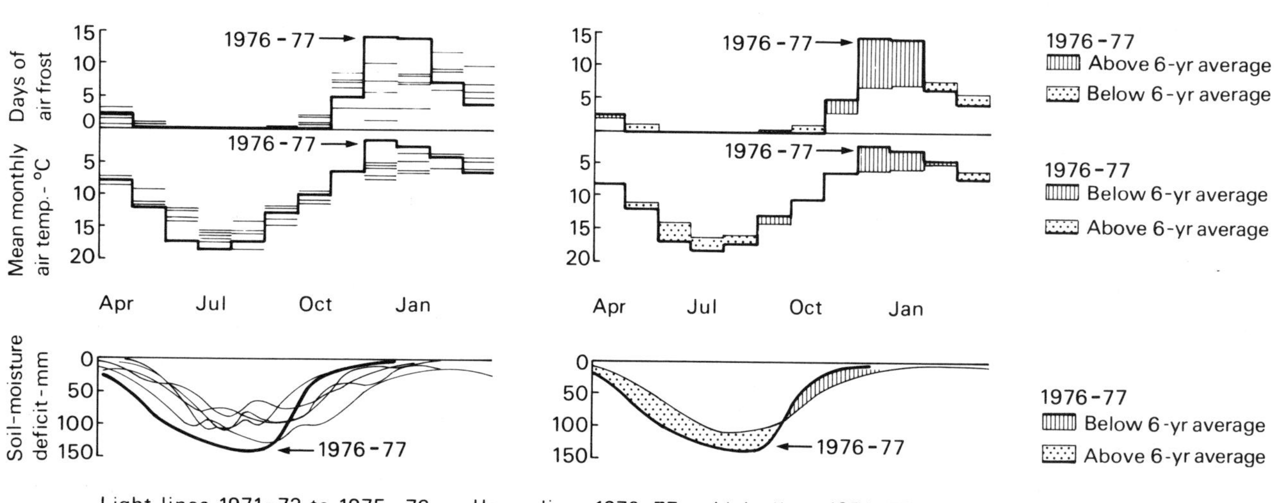


**Figure 18** Mean monthly air temperatures, days of air frost per month, and mean monthly soil-moisture deficits averaged for three sites (Kew, Edgbaston and Edinburgh Royal Botanic Garden): 1971–77 and 1976–77. (Sources: Meteorological Office Monthly Weather Reports and Estimated Soil Moisture Deficit Bulletins)

Further factors that might be expected to affect only the frequency of explosions resulting from external leaks are ground movements in shrinkable clays due to summer drying out by vegetation followed by winter wetting, and winter reductions in the permeability of the ground near the surface. The ground movements are likely to increase the number of mains fractures in the late summer and autumn or early winter in a year when the summer is unusually dry. Winter reductions in the permeability of the ground near the surface will increase the likelihood of gas escaping from a fracture travelling sideways and entering a nearby building (or a duct leading to a building) rather than upwards into the open air. Direct measures of these factors are not available, but the lower plots of Figure 18 show variations in other weather indices to which they are probably related. Unusually large variations in calculated soil-moisture deficit, such as occurred in the late summer and autumn of 1976, must lead to larger ground movements than usual. Also the permeability of the ground near the surface must be considerably reduced when it is saturated with water (ie when the calculated soil-moisture deficit is nil) and it must be further reduced when it is frozen to any significant depth or covered by snow. The number of days of air frost during a month probably gives the best available indication of the relative likelihood of this further reduction. A precise year-by-year correlation between these factors and the actual frequencies of explosions resulting from external leaks should not be expected, both because other factors including random ones are also relevant and because it has been necessary to average the weather indices geographically whereas explosions occur only on specific sites. If some allowance is made for this, the correlation is remarkably good. Risks of gas leakages from broken mains into buildings must again be expected to rise in areas of shrinkable clays when periods of drought are followed by rapid wetting of the ground, and in most areas in periods of heavy frost or snow and when the ground surface is saturated.

### 3.7 The wider context

Few comparable data are available from other countries. Ligtenberg has given brief details of a survey of press reports of all kinds of accidental structural damage in Holland in 1967[18]. There were 200 instances of damage by explosions of all kinds for a total building stock of about 3 x $10^6$. This suggests a much higher risk than the risks estimated in Tables 8a–d, but the risk is probably inflated in comparison with these risks by a broader criterion of structural damage and by the inclusion of a wider range of types of incident. Burnett *et al*[19] have summarised two collections of data for the USA. The more reliable data[20] covered explosions of piped gas only in 1957–66. On the basis of these data it was estimated that 0.042 structurally significant explosions occurred in dwellings per $10^9$ MJ of natural gas sold, the corresponding figure from the present survey for dwellings only being 0.054. Here the criterion of significance – a payment by the Gas Company in excess of $1000 – may well have been, on average, somewhat narrower. There is thus no evidence that the British risks are higher than those in other developed countries.

During the period of the survey some information has been acquired on very severe explosions abroad and a few of these explosions have been inspected by courtesy of the local authorities[8]. The coverage cannot be considered complete, but it is unlikely that many explosions involving structural damage in excess of £100 000 or large numbers of fatalities have been missed in Western Europe. The explosions of this kind of which we know are listed in Table 13. There were two definitely comparable British explosions and one that probably just qualified, but only on account of the unusual difficulty of reinstatement which was not a factor likely to lead to reporting of a similar explosion abroad in the British press. Again, therefore, there is no evidence that the British risks are abnormal.

In two respects however, there has been some underestimation of total risk. Deliberate bombings have been excluded, though these may be considered accidents from the points of view of most building owners and occupiers and their designers, and no account has been taken of damage to non-industrial buildings by industrial explosions. The risk of major explosions of unconfined vapour clouds is probably increasing and their damage potential was well illustrated by the Flixborough explosion (Figure 19)[21–23].

## 4 FURTHER ANALYSIS OF SIGNIFICANT IMPACTS

### 4.1 Risks in relation to type and use of building

Risks of significant vehicle impacts in relation to type and use of building are analysed in Table 14 in the same manner as risks of significant explosions in Tables 8a–d, except that they have been calculated only in terms of numbers of units affected. Because of the undoubted incompleteness of coverage, the risks are certainly underestimated, especially for the later years. The true order of risk is probably best indicated by the annual averages for 1971–73. The later figures are nevertheless of some value in that the relative risks for different types and uses are probably indicated more reliably by the annual averages over all six years. The differences in relativities are, however, small except in the cases of detached houses and sub-stations.

If the relative risks are compared with those for explosions, the chief differences seem to be that bungalows rather than terraced houses are most at risk in the houses and bungalows category, that risks for separate flats are very much less, and that risks for flats over shops and for the other non-industrial categories are also reduced. The highest risks are those for shops, public houses, etc, for flats over these, and for bungalows.

### 4.2 Risks in relation to age and height of building

Risks for dwellings only in relation to age and height of building are analysed in Tables 15 and 16 respectively, again in much the same way as the risks of explosions. Here the correlation between age and risk is even more striking. Dwellings built before 1919 suffered four times as many significant impacts in proportion to their number as dwellings built subsequently on the basis of

**Figure 19** Damage to a three-storey office block with an in-situ reinforced concrete frame by the Flixborough industrial explosion

**Figure 20** Damage resulting from a recent aircraft crash (Reproduced by courtesy of The Times)

the six-year totals, and almost six times as many as dwellings built after 1945. Whereas this variation is in the same sense as that found for explosions, the variation in relation to height of building is in the opposite sense and reflects the relatively higher risk of impact for bungalows and the much lower risk for separate flats – though it is probable that neither height nor type of dwelling is the primary factor.

### 4.3 Other factors influencing the risk

Since impacts all have an external cause, it is not to be expected that type and use of a building would directly affect the risk. It is much more likely that siting in relation to traffic routes is the primary factor. Direct confirmation of this would have entailed a good deal of further enquiry, so it was primarily to seek indirect confirmation that the analyses referred to above were undertaken, and there can be little doubt that they do provide it. A much larger proportion of dwellings built before 1919 than of later dwellings is, for instance, to be found immediately adjacent to main traffic routes. Conversely it is recent tall blocks of flats that are most likely to be set well back from such routes, in addition to which they have, for other reasons, much more strength at the base to resist without harm any impacts that they do suffer. The only slightly surprising finding in this context is the relatively high risk for bungalows. This may partly reflect the relative structural weakness of many bungalows which is likely to lead to significant damage in cases where the damage would not have been considered significant if the structure had been more substantial.

The data do not show any clear seasonal variation in the risk. Adverse winter road conditions might be expected to increase it, but any such increase would probably be counterbalanced by other factors such as increased traffic in the summer. In only about half the incidents were the circumstances that led to the accident reported, but it is notable that about three times as many accidents were caused by driverless or stolen vehicles as were attributed to skidding on wet surfaces or on ice or snow, or to fog.

### 4.4 The wider context

No truly comparable data are available from other countries. The only relevant paper known uses data on 'building collisions' from the States of Oklahoma and Illinois to estimate average risks for the United States[24]. By making some rather arbitrary assumptions it arrives at an annual risk for all dwellings more than an order of magnitude greater than that found in any year in the present survey.

With present volumes of air traffic there is clearly some very small risk of much more serious impacts than have yet been experienced. Some foretaste was perhaps given by a recent accident which destroyed a number of houses (Figure 20), though the damage here was largely the result of an ensuing fire. Quantification of the risk is very difficult and it must be too remote to justify consideration except in the design of particularly vulnerable buildings sited near airports. Somes, on the basis of data supplied by the United States Federal Aviation Authority, has estimated that it is of the order of 0.01 per million buildings per year for buildings located more than 5 km from the end of an airport runway[25]. The fact that there were no such incidents in this country during the six years of the survey suggests that it cannot be higher.

## 5 CONCLUSIONS AND RECOMMENDATIONS

### 5.1 General

1 The risk of accidental loads well in excess of those assumed in normal design continues to be a real one, though there was no further British instance of a consequential progressive collapse on the scale of that at Ronan Point in 1968.

2 If industrial explosions (like that at Flixborough in 1974) and accidents during construction are excluded, such loads arise as a result of other accidental explosions or impacts by road vehicles. They could also arise as a result of impacts by aircraft, but no instance of this was reported during the period covered by this report.

3 On the basis of the press reports received there were, on average, 1.64 non-industrial incidents per week that clearly involved loads of the magnitudes here envisaged and thus led to significant structural damage. This total was made up of 0.76 piped-gas explosions (including 0.30 severe or very severe explosions), 0.23 accidental explosions of other or unknown kinds (including 0.06 severe ones), and 0.65 impacts. Allowing for the incomplete coverage provided by the press reports, the true frequencies were probably between 10 and 20 per cent higher for explosions and higher by a substantially greater margin for impacts. (See Sections 1.5, 1.6 and 2.2.) For comparison with other statistics it should be noted that the estimated costs for only structural reinstatement for these incidents (adjusted to early 1977 prices) were at least £800 for explosions and at least £400 for impacts. (See Section 2.3.)

4 The estimated annual cost of all these non-industrial incidents plus an allowance for less severe ones (but excluding costs of personal injuries and deaths) averaged about £1½ million (again at early 1977 prices). Between 80 and 90 per cent of this was attributable to accidental explosions and the remainder to vehicle impacts. The combined casualties, excluding vehicle drivers and passengers, included at least 206 fatalities over the six years. (Details of those in structurally significant incidents are given in the Appendices.) The inclusive annual cost was probably over £5 million. (See Section 2.3.)

### 5.2 Significant accidental explosions

1 Frequencies of significant explosions varied considerably with type and use of building from 1.5 per million units per year for detached houses to about 10 times this figure for a group consisting chiefly of public utility sub-stations and telephone exchanges. Within the broad category of all dwellings there were progressively higher frequencies for semi-detached houses, terraced houses, flats and bungalows with an overall average of 2.1 per

million units per year. In multi-occupation or multi-use buildings, a single explosion often caused significant damage to more than one occupancy unit. Frequencies in terms of numbers of units affected varied between the same overall limits as the frequencies of explosions, but they were much higher than these frequencies for flats, shops, etc, and flats over shops, being 4.3, 10.9 and 13.6 per million units per year respectively for 2.3 explosions per million units per year in flats and 5.7 in shops. (See Section 3.2.)

2 There were smaller variations in frequency in relation to age and height of building, the frequencies being higher for older and taller buildings. (See Sections 3.3 and 3.4.)

3 By far the largest number of explosions was of piped gas. Together with explosions of liquefied (cylinder) gas they accounted for at least 87 per cent of the total. They alone accounted for all very severe explosions. In relation to consumption, however, piped gas was found to be a much safer fuel than liquefied (cylinder) gas. The risk of explosion with the latter was at least an order of magnitude greater in terms of energy consumed. (See Section 3.5.)

4 Conversion from town (manufactured) gas to natural gas has reduced the risk of explosion in terms of energy consumed. For significant explosions in dwellings the reduction has been from about 0.08 per $10^9$ MJ of town gas consumed to about 0.06 per $10^9$ MJ of natural gas consumed. (See Section 3.5.)

5 Conversion has, however, led to some increase in the frequency of explosions where the source of gas was a leaking main or service pipe: all the reduction has been in the frequency of explosions where the source was an appliance within the building or leaking installation pipework, and may have been partly due to the testing of installations that accompanied the conversion process and a changing pattern of consumption. (See Section 3.5.)

6 Explosions where gas leaked into the building from outside therefore now account for a much larger proportion of the total of significant piped-gas explosions. Over the six years, 48 per cent of natural-gas explosions were of this kind, including 57 per cent of those classified as severe or worse and 73 per cent of those classified as very severe. The occurrence of such explosions has been found to be closely correlated with certain aspects of the weather. Large variations in soil-moisture deficit, leading to ground movements in shrinkable clays, are one likely cause of breaks in mains and service pipes. More importantly, very cold weather, especially when the ground is saturated or covered in snow, seems to lead to escaping gas travelling sideways into nearby buildings more readily than in dry warm weather when it can easily escape upwards into the open air. A pronounced peak in explosions of this kind in December 1976 occurred, for instance, in unusually cold weather after a summer and autumn with an extreme variation in soil-moisture content. (See Sections 3.5 and 3.6.)

7 Estimates of peak pressures reached in explosions of piped gas, and made on the basis of inspections of the resultant damage, suggest that an equivalent static pressure of about 34 kN/m$^2$ may be reached in about one accidental explosion a year. There may be about five explosions a year in which an equivalent static pressure of half this magnitude is reached or exceeded. (See Section 3.1.)

8 Similar estimates made for 17 explosions of natural gas and 22 broadly comparable ones of town gas indicated average peak pressures in the former of 11 kN/m$^2$ and in the latter of 13 kN/m$^2$, though it is not considered that this comparison is as meaningful as those between frequencies of explosions of given severities referred to above. (See Section 3.1 and Table 7.)

### 5.3 Significant impacts

1 Frequencies of occurrence of significant impacts also varied widely with use, type, age and height of building, though the true risks were much less precisely established because of shortcomings in the press coverage. The frequencies were greatest for shops, public houses and similar buildings, for flats over these, and for bungalows. For dwellings built before 1919 the average frequency was six times as high as for those built after 1945. (See Sections 4.1 and 4.2.)

2 It is, however, considered that these variations are only indirectly related to the variables considered, and not directly as in the case of many explosions. They almost certainly arise primarily because of the different siting of buildings falling in the different categories in relation to traffic routes. (See Section 4.3.)

### 5.4 Recommendations

1 The additional load of 34 kN/m$^2$ specified in paragraph (5) of Building Regulation D17 (1976) should be retained if the cost of meeting the requirement is considered reasonable for averting the risk of about one serious partial collapse of a building of five or more storeys once every 25 years (assuming that 4 per cent of significant explosions continue to occur in buildings of this height). If the additional load were reduced to half this figure, a risk would remain of about one such collapse every five years.

2 There should be no relaxation of this requirement for buildings without a piped gas supply, since a risk would usually remain of gas leaking into the building from outside and a higher risk would be incurred if cylinders of liquefied gas were introduced to even a much more limited extent.

3 Efforts at reducing the risk of damaging explosions of piped gas should be concentrated now on the risk arising from leaks external to the building, and should pay particular attention to the higher-than-average risks in very cold weather. The sealing of possible entry points to a building and the maintenance of good ventilation, particularly below ground level, are both clearly desirable (bearing in mind that possible entry points

include ducts carrying other services). Attention should also continue to be given to reducing the basic risk of leaks from underground pipes.

4 Users of cylinders of liquefied gas should be made more aware of the risks arising from improper use and faulty equipment, and should be better instructed in correct procedures.

5 Though some buildings are more likely to be damaged by impacts of road vehicles than by accidental explosions, this seems to be largely on account of their siting in relation to traffic routes. As they are mostly old buildings that it would be impracticable or uneconomic to strengthen usefully, reduction of the risk of impact damage calls chiefly for re-routing of heavy traffic away from narrow streets lined by old buildings, though protective bollards, curbs or low walls may be a reasonable alternative in some cases. New buildings should be sited away from heavy traffic, or protected in similar ways and designed without vulnerable elements on the perimeter at ground level.

6 The risk of impacts from aircraft seems to be too small in general to call for any special provisions in design.

7 Further recommendations for the structural design of buildings, particularly those of less than five storeys, to withstand better the effects of accidental loads have been made separately on the basis of detailed examinations of the damage caused in particular incidents[26].

## 6 REFERENCES


1 **Griffiths H, Pugsley Sir A and Saunders Sir O.** Report of the inquiry into the collapse of flats at Ronan Point, Canning Town. London, HMSO, 1968.

2 The Building Regulations 1976. Statutory Instrument 1976 No 1676 Building and Buildings. London, HMSO, 1976.

3 **British Standards Institution.** The structural use of concrete. BS Code of Practice CP 110:1972. London, BSI, 1972.

4 **Department of the Environment.** Structural Stability. Research Report 3. London, DOE, 1975.

5 **King P J, Clegg G T and Walters W J.** Report of the inquiry into serious gas explosions. London, HMSO, 1977.

6 **Taylor N and Alexander S J.** Structural damage in buildings caused by gaseous explosions and other accidental loadings. Building Research Establishment Current Paper CP45/74. Garston, BRE, 1974.

7 **Mainstone R J and Butlin R N.** Report on an explosion at Mersey House, Bootle, Lancs. Building Research Establishment Current Paper CP34/76. Garston, BRE, 1976.

8 **Mainstone R J.** The response of buildings to accidental explosions. Conference on performance of building structures, Glasgow, 1976, Vol 1, pp 173–188. London, Pentech Press, 1976. (Also Building Research Establishment Current Paper CP24/76. Garston, BRE, 1976.)

9 **Mainstone R J** (Editor). Buildings and the Hazard of Explosion. Proceedings of a symposium at the Building Research Establishment, Garston, 18 October 1972. Garston, BRE, 1974.

10 **Mainstone R J.** The hazards of explosion, impact, and other random loadings on tall buildings. Tall Buildings and People? IABSE/ISE Conference, Oxford, 1974, pp 122–129. London, ISE, 1974. (Also Building Research Establishment Current Paper CP64/74. Garston, BRE, 1974.)

11 **Mainstone R J.** The breakage of glass windows by gas explosions. Building Research Establishment Current Paper CP26/71. Garston, BRE, 1971.

12 **Astbury N F, West H W H, Hodgkinson H R, Cubbage P A and Clare R.** Gas explosions in loadbearing brick structures. Special Publication No 68. Stoke-on-Trent, BCRA, 1970.

13 **Rasbash D J.** Gas explosions in multiple compartments. Fire Research Note No 847. Borehamwood, FRS, 1970. (Also London, DOE, 1971.)

14 **Dragosavic M.** Structural measures against natural-gas explosions in high-rise blocks of flats. Heron, 1973, **19** (4) 1–51.

15 **Butlin R N and Tonkin P S.** Pressures produced by gas explosions in a vented compartment. Fire Research Note No 1019. Borehamwood, FRS, 1974.

16 on the basis of the construction cost indices given in: Housing and Construction Statistics. London, HMSO, quarterly.

17 **Morton F.** Report of the inquiry into the safety of natural gas as a fuel. London, HMSO, 1970.

18 **Ligtenberg F K.** Structural safety and catastrophic events. Symposium on concepts of safety of structures and methods of design, London, 1969. Final Report, pp 25–33. Zurich, IABSE, 1969.

19 **Burnett E F P, Somes N F and Leyendecker E V.** Residential buildings and gas-related explosions. NBSIR 73–208. Washington, NBS, 1973.

20 **Little A D**. Public safety and gas distribution. Survey for the American Gas Association, 1967.

21 **Parker R J** et al. The Flixborough disaster. Report of the court of inquiry. London, HMSO, 1975.

22 **Strehlow R A**. Unconfined vapor-cloud explosions – an overview. Fourteenth symposium on combustion, Pennsylvania State University, 1972, pp 1189–1200. Pittsburgh, The Combustion Institute, 1973.

23 **Steel W A, Bowser D and Chapman R E**. The incidence of hazardous material accidents during transportation and storage. NBSIR 73–412. Washington, NBS, 1973.

24 **Fribush S L, Bowser D and Chapman R**. Estimates of vehicular collisions with multistorey buildings. NBSIR 73–175. Washington, NBS, 1973.

25 **Somes N F**. Abnormal loading of buildings and progressive collapse. NBSIR 73–221. Washington, NBS, 1973.

26 **Mainstone R J**. Accidental explosions and impacts: some lessons from recent incidents. Institution of Structural Engineers Symposium on Low-Rise Construction, July 1978.

## TABLES

**Table 1 Comparison between calendar-year totals of reported and significant piped-gas explosions in dwellings from Fire Research Station/Home Office data and from the present survey**

| Calendar year | Sampling fraction | Reported incidents | | Significant incidents | |
|---|---|---|---|---|---|
| | | Fire Research Station | Survey | Fire Research Station | Survey |
| 1957 | 1/1 | 69 | | 8 | |
| 1958 | 1/4 | 60 | | 4 | |
| 1959 | 1/4 | 84 | | 28 | |
| 1960 | 1/4 | 72 | | 20 | |
| 1961 | 1/2 | 46 | | 8 | |
| 1962 | 1/2 | 68 | | 4 | |
| 1963 | 1/6 | 84 | | 0 | |
| 1964 | 1/2 | 78 | | 24 | |
| 1965 | 1/1 | 75 | | 14 | |
| 1966 | 1/1 | 96 | | 22 | |
| 1967 | 1/1 | 104 | | 12 | |
| 1968 | 1/4 | 152 | | 20 | |
| 1969 | 1/1 | 156 | | 34 | |
| 1970 | 1/1 | 188 | | – | |
| 1971 | 1/1 | 189 | 153* | – | 28* (33)* |
| 1972 | 1/1 | 213 | 193 | 65 | 38 (39) |
| 1973 | 1/2 | 213 | 166 | 44 | 27 (28) |
| 1974 | 1/1 | 200 | 162 | 66 | 35 (37) |
| 1975 | ** | – | 184 | – | 27 (28) |
| 1976 | | | 175 | | 40 (41) |

*Extrapolated to full year from 39 weeks' records April–December
**Returns to Fire Research Station for this year incomplete

The sampling fraction refers to the analysis of the Fire Research Station's data
Figures in parentheses include incidents in which a dwelling was damaged by an explosion in a shop in the same building

**Table 2 Comparison between survey-year (April–March) totals of significant and severe piped-gas explosions in all non-industrial buildings and of piped-gas explosions causing fatalities from British Gas data and from the present survey**

| Survey year | 'Significant' explosions | | 'Severe' explosions | | Explosions causing fatalities | |
|---|---|---|---|---|---|---|
| | Survey | British Gas* | Survey | British Gas* | Survey† | British Gas |
| 1971–72 | 51 | | 22 | | 10 (34) | |
| 1972–73 | 36 | 104 | 18 | 26 | 8 ( 8) | 8 ( 8) |
| 1973–74 | 32 | 98 | 12 | 26 | 8 ( 9) | 10 (11) |
| 1974–75 | 38 | 100 | 14 | 28 | 15 (17) | 14 (17) |
| 1975–76 | 37 | 128 | 9 | 27 | 12 (13) | 10 (11) |
| 1976–77 | 43 | 111 | 19 | 37 | 10 (10) | 9 (10) |

*These totals probably include small, but unknown,numbers of explosions in industrial buildings
†These totals exclude incidents in which the fatalities were primarily due to burns or suffocation during an associated fire
Figures in parentheses are numbers of known fatalities

**Table 3a Quarterly and annual totals of all reported incidents broken down in terms of type of incident: 1971–73**

| Type of incident | Code | 1971–72 | | | | | 1972–73 | | | | | Total 1971–73 (2 years) |
|---|---|---|---|---|---|---|---|---|---|---|---|---|
| | | April–June | July–Sept | Oct–Dec | Jan–March | Year | April–June | July–Sept | Oct–Dec | Jan–March | Year | |
| Explosions | | | | | | | | | | | | |
| Town gas | T | 30 | 23 | 37 | 50 | 140 | 27 | 18 | 35 | 48 | 128 | 268 |
| Natural gas | N | 17 | 18 | 33 | 70 | 138 | 31 | 34 | 35 | 65 | 165 | 303 |
| Piped LPG | LPG | 0 | 0 | 0 | 0 | 0 | 0 | 0 | 0 | 0 | 0 | 0 |
| Total piped gas | | 47 | 41 | 70 | 120 | 278 | 58 | 52 | 70 | 113 | 293 | 571 |
| Liquefied gas (cylinders, aerosols) | C | 23 | 10 | 32 | 70 | 135 | 33 | 26 | 23 | 54 | 136 | 271 |
| Bombs, explosives | B | 13 | 10 | 18 | 19 | 60 | 11 | 18 | 15 | 13 | 57 | 117 |
| Other materials (dust, vapour, chemicals) | P/S/O | 27 | 21 | 41 | 56 | 145 | 47 | 37 | 33 | 34 | 151 | 296 |
| Water vapour | W | 3 | 4 | 5 | 7 | 19 | 5 | 5 | 6 | 6 | 22 | 41 |
| Miscellaneous causes (including lightning) | X | 21 | 4 | 42 | 38 | 105 | 17 | 18 | 13 | 22 | 70 | 175 |
| Total other known causes | | 87 | 49 | 138 | 190 | 464 | 113 | 104 | 90 | 129 | 436 | 900 |
| Unknown causes | U | 29 | 42 | 35 | 50 | 156 | 42 | 28 | 38 | 49 | 157 | 313 |
| Total, all causes | | 163 | 132 | 243 | 360 | 898 | 213 | 184 | 198 | 291 | 886 | 1784 |
| Vehicle impacts | I | 29 | 47 | 42 | 68 | 186 | 64 | 66 | 64 | 65 | 259 | 445 |

**Table 3b Quarterly and annual totals of all reported incidents broken down in terms of type of incident: 1973–75**

| Type of incident | Code | 1973–74 | | | | | 1974–75 | | | | | Total 1973–75 (2 years) |
|---|---|---|---|---|---|---|---|---|---|---|---|---|
| | | April–June | July–Sept | Oct–Dec | Jan–March | Year | April–June | July–Sept | Oct–Dec | Jan–March | Year | |
| Explosions | | | | | | | | | | | | |
| Town gas | T | 20 | 10 | 22 | 20 | 72 | 13 | 16 | 7 | 6 | 42 | 114 |
| Natural gas | N | 26 | 23 | 37 | 33 | 119 | 48 | 32 | 48 | 65 | 193 | 312 |
| Piped LPG | LPG | 0 | 0 | 0 | 0 | 0 | 0 | 0 | 0 | 0 | 0 | 0 |
| Total piped gas | | 46 | 33 | 59 | 53 | 191 | 61 | 48 | 55 | 71 | 235 | 426 |
| Liquefied gas (cylinders, aerosols) | C | 42 | 25 | 28 | 42 | 137 | 25 | 28 | 45 | 26 | 124 | 261 |
| Bombs, explosives | B | 9 | 7 | 1 | 0 | 17 | 2 | 15 | 25 | 8 | 50 | 67 |
| Other materials (dust, vapour, chemicals) | P/S/O | 47 | 24 | 34 | 27 | 132 | 38 | 38 | 58 | 43 | 177 | 309 |
| Water vapour | W | 5 | 5 | 5 | 3 | 18 | 8 | 3 | 10 | 5 | 26 | 44 |
| Miscellaneous causes (including lightning) | X | 23 | 11 | 11 | 14 | 59 | 16 | 12 | 17 | 16 | 61 | 120 |
| Total other known causes | | 126 | 72 | 79 | 86 | 363 | 89 | 96 | 155 | 98 | 438 | 801 |
| Unknown causes | U | 31 | 22 | 11 | 13 | 77 | 26 | 27 | 20 | 20 | 93 | 170 |
| Total, all causes | | 203 | 127 | 149 | 152 | 631 | 176 | 171 | 230 | 189 | 766 | 1397 |
| Vehicle impact | I | 16 | 24 | 17 | 8 | 65 | 18 | 24 | 11 | 9 | 62 | 127 |

**Table 3c Quarterly and annual totals of all reported incidents broken down in terms of type of incident: 1975–77**

| Type of incident | Code | 1975–76 | | | | | 1976–77 | | | | | Total 1975–77 (2 years) |
|---|---|---|---|---|---|---|---|---|---|---|---|---|
| | | April–June | July–Sept | Oct–Dec | Jan–March | Year | April–June | July–Sept | Oct–Dec | Jan–March | Year | |
| Explosions | | | | | | | | | | | | |
| Town gas | T | 1 | 3 | 10 | 2 | 16 | 1 | 1 | 0 | 2 | 4 | 20 |
| Natural gas | N | 43 | 52 | 64 | 85 | 244 | 50 | 48 | 59 | 106 | 263 | 507 |
| Piped LPG | LPG | 0 | 0 | 0 | 0 | 0 | 0 | 0 | 1 | 0 | 1 | 1 |
| Total piped gas | | 44 | 55 | 74 | 87 | 260 | 51 | 49 | 60 | 108 | 268 | 528 |
| Liquefied gas (cylinders, aerosols) | C | 43 | 46 | 67 | 59 | 215 | 47 | 50 | 36 | 67 | 200 | 415 |
| Bombs, explosives | B | 5 | 8 | 5 | 12 | 30 | 7 | 5 | 2 | 10 | 24 | 54 |
| Other materials (dust,vapour, chemicals) | P/S/O | 39 | 33 | 40 | 81 | 193 | 71 | 56 | 51 | 44 | 222 | 415 |
| Water vapour | W | 3 | 5 | 4 | 6 | 18 | 5 | 3 | 4 | 3 | 15 | 33 |
| Miscellaneous causes (including lightning) | X | 8 | 8 | 18 | 24 | 58 | 20 | 19 | 34 | 30 | 103 | 161 |
| Total other known causes | | 98 | 100 | 134 | 182 | 514 | 150 | 133 | 127 | 154 | 564 | 1078 |
| Unknown causes | U | 13 | 27 | 24 | 23 | 87 | 35 | 28 | 25 | 27 | 115 | 202 |
| Total, all causes | | 155 | 182 | 232 | 292 | 861 | 236 | 210 | 212 | 289 | 947 | 1808 |
| Vehicle impacts | I | 7 | 7 | 14 | 26 | 54 | 20 | 27 | 38 | 49 | 134 | 188 |

**Table 4a Further analysis of all reported incidents in terms of use and type of building: 1971–73**

| Type of incident | Code | 2-storey dwellings and unspecified | 1-storey dwellings | Dwellings of 3 storeys and above | Offices | Shops, cafes, public houses | Hospitals | Educational buildings | Other buildings (non-industrial) | Temporary buildings including caravans | Electrical sub-stations | External | Total non-industrial | Total industrial | Total all incidents 1971–73 (2 years) |
|---|---|---|---|---|---|---|---|---|---|---|---|---|---|---|---|
| Explosions | | | | | | | | | | | | | | | |
| Town gas | T | 156 | 5 | 14 | 3 | 26 | 2 | 1 | 4 | 1 | 0 | 39 | 251 | 17 | 268 |
| Natural gas | N | 191 | 12 | 8 | 8 | 15 | 1 | 6 | 2 | 1 | 1 | 40 | 285 | 18 | 303 |
| Piped LPG | LPG | 0 | 0 | 0 | 0 | 0 | 0 | 0 | 0 | 0 | 0 | 0 | 0 | 0 | 0 |
| Total piped gas | | 347 | 17 | 22 | 11 | 41 | 3 | 7 | 6 | 2 | 1 | 79 | 536 | 35 | 571 |
| Liquefied gas (cylinders, aerosols) | C | 102 | 6 | 1 | 10 | 15 | 2 | 2 | 3 | 17 | 1 | 79 | 238 | 33 | 271 |
| Bombs, explosives | B | 23 | 1 | 6 | 4 | 10 | 0 | 5 | 7 | 1 | 1 | 46 | 104 | 13 | 117 |
| Other materials (dust, vapour, chemicals) | P/S/O | 79 | 6 | 4 | 6 | 7 | 1 | 17 | 7 | 1 | 3 | 40 | 171 | 125 | 296 |
| Water vapour | W | 25 | 0 | 1 | 2 | 1 | 1 | 2 | 1 | 0 | 0 | 0 | 33 | 8 | 41 |
| Miscellaneous causes (including lightning) | X | 57 | 8 | 8 | 0 | 22 | 4 | 6 | 2 | 1 | 0 | 34 | 142 | 33 | 175 |
| Total other known causes | | 286 | 21 | 20 | 22 | 55 | 8 | 32 | 20 | 20 | 5 | 199 | 688 | 212 | 900 |
| Unknown causes | U | 65 | 10 | 4 | 5 | 20 | 2 | 7 | 2 | 2 | 85 | 42 | 244 | 69 | 313 |
| Total, all causes | | 698 | 48 | 46 | 38 | 116 | 13 | 46 | 28 | 24 | 91 | 320 | 1468 | 316 | 1784 |
| Vehicle impacts | I | 199 | 21 | 2 | 5 | 100 | 2 | 6 | 7 | 0 | 0 | 100 | 442 | 3 | 445 |

**Table 4b Further analysis of all reported incidents in terms of use and type of building: 1973–75**

| Type of incident | Code | 2-storey dwellings and unspecified | 1-storey dwellings | Dwellings of 3 storeys and above | Offices | Shops, cafes, public houses | Hospitals | Educational buildings | Other buildings (non-industrial) | Temporary buildings including caravans | Electrical sub-stations | External | Total non-industrial | Total industrial | Total all incidents 1973–75 (2 years) |
|---|---|---|---|---|---|---|---|---|---|---|---|---|---|---|---|
| Explosions | | | | | | | | | | | | | | | |
| Town gas | T | 72 | 2 | 2 | 3 | 5 | 0 | 2 | 4 | 0 | 0 | 15 | 105 | 9 | 114 |
| Natural gas | N | 213 | 7 | 5 | 3 | 29 | 2 | 8 | 14 | 0 | 1 | 21 | 303 | 9 | 312 |
| Piped LPG | LPG | 0 | 0 | 0 | 0 | 0 | 0 | 0 | 0 | 0 | 0 | 0 | 0 | 0 | 0 |
| Total piped gas | | 285 | 9 | 7 | 6 | 34 | 2 | 10 | 18 | 0 | 1 | 36 | 408 | 18 | 426 |
| Liquefied gas (cylinders, aerosols) | C | 65 | 5 | 3 | 7 | 26 | 4 | 4 | 16 | 74 | 0 | 21 | 225 | 36 | 261 |
| Bombs, explosives | B | 5 | 0 | 0 | 4 | 10 | 0 | 2 | 13 | 3 | 0 | 23 | 60 | 7 | 67 |
| Other materials (dust, vapour, chemicals) | P/S/O | 39 | 0 | 4 | 1 | 10 | 4 | 13 | 29 | 10 | 2 | 27 | 139 | 170 | 309 |
| Water vapour | W | 27 | 0 | 0 | 1 | 0 | 3 | 2 | 0 | 1 | 2 | 0 | 36 | 8 | 44 |
| Miscellaneous causes (including lightning) | X | 54 | 1 | 1 | 1 | 8 | 0 | 2 | 6 | 3 | 0 | 20 | 96 | 24 | 120 |
| Total other known causes | | 190 | 6 | 8 | 14 | 54 | 11 | 23 | 64 | 91 | 4 | 91 | 556 | 245 | 801 |
| Unknown causes | U | 49 | 4 | 2 | 0 | 15 | 1 | 4 | 15 | 6 | 28 | 17 | 141 | 29 | 170 |
| Total, all causes | | 524 | 19 | 17 | 20 | 103 | 14 | 37 | 97 | 97 | 33 | 144 | 1105 | 292 | 1397 |
| Vehicle impacts | I | 67 | 2 | 0 | 1 | 31 | 0 | 0 | 1 | 0 | 0 | 24 | 126 | 1 | 127 |

**Table 4c Further analysis of all reported incidents in terms of use and type of building: 1975–77**

| Type of incident | Code | 2-storey dwellings and unspecified | 1-storey dwellings | Dwellings of 3 storeys and above | Offices | Shops, cafes, public houses | Hospitals | Educational buildings | Other buildings (non-industrial) | Temporary buildings including caravans | Electrical sub-stations | External | Total non-industrial | Total industrial | Total all incidents 1975–77 (2 years) |
|---|---|---|---|---|---|---|---|---|---|---|---|---|---|---|---|
| Explosions | | | | | | | | | | | | | | | |
| Town gas | T | 12 | 0 | 1 | 0 | 1 | 1 | 0 | 2 | 0 | 0 | 1 | 18 | 2 | 20 |
| Natural gas | N | 347 | 16 | 17 | 5 | 29 | 1 | 12 | 19 | 0 | 0 | 47 | 493 | 14 | 507 |
| Piped gas | LPG | 1 | 0 | 0 | 0 | 0 | 0 | 0 | 0 | 0 | 0 | 0 | 1 | 0 | 1 |
| Total piped gas | | 360 | 16 | 18 | 5 | 30 | 2 | 12 | 21 | 0 | 0 | 48 | 512 | 16 | 528 |
| Liquefied gas (cylinders, aerosols) | C | 120 | 12 | 2 | 7 | 35 | 1 | 2 | 22 | 120 | 0 | 60 | 381 | 34 | 415 |
| Bombs, explosives | B | 18 | 0 | 1 | 1 | 2 | 0 | 1 | 2 | 3 | 0 | 21 | 49 | 5 | 54 |
| Other materials (dust, vapour, chemicals) | P/S/O | 64 | 4 | 2 | 2 | 25 | 2 | 21 | 32 | 14 | 0 | 38 | 204 | 211 | 415 |
| Water vapour | W | 17 | 1 | 0 | 2 | 3 | 1 | 1 | 1 | 0 | 0 | 0 | 26 | 7 | 33 |
| Miscellaneous causes (including lightning) | X | 47 | 4 | 1 | 6 | 12 | 3 | 3 | 10 | 1 | 0 | 41 | 128 | 33 | 161 |
| Total other known causes | | 266 | 21 | 6 | 18 | 77 | 7 | 28 | 67 | 138 | 0 | 160 | 788 | 290 | 1078 |
| Unknown causes | U | 42 | 3 | 9 | 3 | 8 | 2 | 3 | 9 | 6 | 76 | 20 | 181 | 21 | 202 |
| Total, all causes | | 668 | 40 | 33 | 26 | 115 | 11 | 43 | 97 | 144 | 76 | 228 | 1481 | 327 | 1808 |
| Vehicle impacts | I | 84 | 5 | 0 | 0 | 51 | 0 | 1 | 11 | 1 | 0 | 34 | 187 | 1 | 188 |

## Table 4d Further analysis of all reported incidents in terms of use and type of building: 1971–77

| Type of incident | Code | 2-storey dwellings and unspecified | 1-storey dwellings | Dwellings of 3 storeys and above | Offices | Shops, cafes, public houses | Hospitals | Educational buildings | Other buildings (non-industrial) | Temporary buildings including caravans | Electrical sub-stations | External | Total non-industrial | Total industrial | Total all incidents 1971–77 (6 years) |
|---|---|---|---|---|---|---|---|---|---|---|---|---|---|---|---|
| Explosions | | | | | | | | | | | | | | | |
| Town gas | T | 240 | 7 | 17 | 6 | 32 | 3 | 3 | 10 | 1 | 0 | 55 | 374 | 28 | 402 |
| Natural gas | N | 751 | 35 | 30 | 16 | 73 | 4 | 26 | 35 | 1 | 2 | 108 | 1081 | 41 | 1122 |
| Piped LPG | LPG | 1 | 0 | 0 | 0 | 0 | 0 | 0 | 0 | 0 | 0 | 0 | 1 | 0 | 1 |
| Total piped gas | | 992 | 42 | 47 | 22 | 105 | 7 | 29 | 45 | 2 | 2 | 163 | 1456 | 69 | 1525 |
| Liquefied gas (cylinders, aerosols) | C | 287 | 23 | 6 | 24 | 76 | 7 | 8 | 41 | 211 | 1 | 160 | 844 | 103 | 947 |
| Bombs, explosives | B | 46 | 1 | 7 | 9 | 22 | 0 | 8 | 22 | 7 | 1 | 90 | 213 | 25 | 238 |
| Other materials (dust, vapour, chemicals) | P/S/O | 182 | 10 | 10 | 9 | 42 | 7 | 51 | 68 | 25 | 5 | 105 | 514 | 506 | 1020 |
| Water vapour | W | 69 | 1 | 1 | 5 | 4 | 5 | 5 | 2 | 1 | 2 | 0 | 95 | 23 | 118 |
| Miscellaneous causes (including lightning) | X | 158 | 13 | 10 | 7 | 42 | 7 | 11 | 18 | 5 | 0 | 95 | 366 | 90 | 456 |
| Total other known causes | | 742 | 48 | 34 | 54 | 186 | 26 | 83 | 151 | 249 | 9 | 450 | 2032 | 747 | 2779 |
| Unkown causes | U | 156 | 17 | 15 | 8 | 43 | 5 | 14 | 26 | 14 | 189 | 79 | 566 | 119 | 685 |
| Total, all causes | | 1890 | 107 | 96 | 84 | 334 | 38 | 126 | 222 | 265 | 200 | 692 | 4054 | 935 | 4989 |
| Vehicle impacts | I | 350 | 28 | 2 | 6 | 182 | 2 | 7 | 19 | 1 | 0 | 158 | 755 | 5 | 760 |

**Table 5a Quarterly and annual totals of all significant incidents broken down in terms of type of incident: 1971–73**

| Type of incident | Code | 1971–72 | | | | | 1972–73 | | | | | Total 1971–73 (2 years) |
|---|---|---|---|---|---|---|---|---|---|---|---|---|
| | | April–June | July–Sept | Oct–Dec | Jan–March | Year | April–June | July–Sept | Oct–Dec | Jan–March | Year | |
| Explosions | | | | | | | | | | | | |
| Town gas | T | 6 | 7 | 7 | 11 | 31 | 2 | 6 | 2 | 5 | 15 | 46 |
| Natural gas | N | 3 | 5 | 2 | 10 | 20 | 6 | 2 | 8 | 5 | 21 | 41 |
| Piped LPG | LPG | 0 | 0 | 0 | 0 | 0 | 0 | 0 | 0 | 0 | 0 | 0 |
| Total piped gas | | 9 | 12 | 9 | 21 | 51 | 8 | 8 | 10 | 10 | 36 | 87 |
| Liquefied gas (cylinders, aerosols) | C | 0 | 3 | 0 | 1 | 4 | 0 | 0 | 1 | 0 | 1 | 5 |
| Other materials (dust, vapour chemicals) | P/S/O | 1 | 1 | 2 | 2 | 6 | 1 | 0 | 1 | 0 | 2 | 8 |
| Water vapour | W | 0 | 2 | 1 | 3 | 6 | 0 | 0 | 1 | 0 | 1 | 7 |
| Lightning | X | 0 | 0 | 0 | 0 | 0 | 0 | 0 | 0 | 0 | 0 | 0 |
| Total other known causes | | 1 | 6 | 3 | 6 | 16 | 1 | 0 | 3 | 0 | 4 | 20 |
| Unknown causes | U | 0 | 0 | 0 | 0 | 0 | 0 | 0 | 1 | 0 | 1 | 1 |
| Total, all causes | | 10 | 18 | 12 | 27 | 67 | 9 | 8 | 14 | 10 | 41 | 108 |
| Vehicle impacts | I | 10 | 21 | 20 | 17 | 68 | 12 | 15 | 13 | 11 | 51 | 119 |

**Table 5b Quarterly and annual totals of all significant incidents broken down in terms of type of incident: 1973–75**

| Type of incident | Code | 1973–74 | | | | | 1974–75 | | | | | Total 1973–75 (2 years) |
|---|---|---|---|---|---|---|---|---|---|---|---|---|
| | | April–June | July–Sept | Oct–Dec | Jan–March | Year | April–June | July–Sept | Oct–Dec | Jan–March | Year | |
| Explosions | | | | | | | | | | | | |
| Town gas | T | 1 | 1 | 4 | 0 | 6 | 2 | 5 | 0 | 1 | 8 | 14 |
| Natural gas | N | 4 | 3 | 8 | 11 | 26 | 6 | 6 | 9 | 9 | 30 | 56 |
| Piped LPG | LPG | 0 | 0 | 0 | 0 | 0 | 0 | 0 | 0 | 0 | 0 | 0 |
| Total piped gas | | 5 | 4 | 12 | 11 | 32 | 8 | 11 | 9 | 10 | 38 | 70 |
| Liquefied gas (cylinders, aerosols) | C | 3 | 2 | 1 | 1 | 7 | 1 | 3 | 3 | 0 | 7 | 14 |
| Other materials (dust, vapour chemicals) | P/S/O | 1 | 1 | 0 | 0 | 2 | 0 | 0 | 0 | 1 | 1 | 3 |
| Water vapour | W | 1 | 1 | 1 | 0 | 3 | 0 | 0 | 1 | 1 | 2 | 5 |
| Lightning | X | 0 | 0 | 0 | 0 | 0 | 1 | 0 | 0 | 0 | 1 | 1 |
| Total other known causes | | 5 | 4 | 2 | 1 | 12 | 2 | 3 | 4 | 2 | 1 | 23 |
| Unkown causes | U | 0 | 0 | 1 | 0 | 1 | 2 | 0 | 0 | 0 | 2 | 3 |
| Total, all causes | | 10 | 8 | 15 | 12 | 45 | 12 | 14 | 13 | 12 | 51 | 96 |
| Vehicle impacts | I | 5 | 8 | 5 | 2 | 20 | 7 | 7 | 1 | 4 | 19 | 39 |

**Table 5c Quarterly and annual totals of all significant incidents broken down in terms of type of incident: 1975–77**

| Type of incident | Code | 1975–76 | | | | | 1976–77 | | | | | Total 1975–77 (2 years) |
|---|---|---|---|---|---|---|---|---|---|---|---|---|
| | | April–June | July–Sept | Oct–Dec | Jan–March | Year | April–June | July–Sept | Oct–Dec | Jan–March | Year | |
| Explosions | | | | | | | | | | | | |
| Town gas | T | 0 | 0 | 1 | 0 | 1 | 0 | 0 | 0 | 1 | 1 | 2 |
| Natural gas | N | 5 | 3 | 12 | 16 | 36 | 6 | 7 | 13 | 15 | 41 | 77 |
| Piped LPG | LPG | 0 | 0 | 0 | 0 | 0 | 0 | 0 | 1 | 0 | 1 | 1 |
| Total piped gas | | 5 | 3 | 13 | 16 | 37 | 6 | 7 | 14 | 16 | 43 | 80 |
| Liquefied gas (cylinders, aerosols) | C | 2 | 1 | 1 | 2 | 6 | 2 | 1 | 3 | 2 | 8 | 14 |
| Other materials (dust, vapour, chemicals) | P/S/O | 0 | 0 | 0 | 0 | 0 | 0 | 0 | 0 | 1 | 1 | 1 |
| Water vapour | W | 0 | 1 | 0 | 1 | 2 | 1 | 1 | 0 | 0 | 2 | 4 |
| Lightning | X | 0 | 0 | 0 | 0 | 0 | 1 | 0 | 0 | 0 | 1 | 1 |
| Total other known causes | | 2 | 2 | 1 | 3 | 8 | 4 | 2 | 3 | 3 | 12 | 20 |
| Unknown causes | U | 1 | 0 | 0 | 0 | 1 | 1 | 1 | 1 | 1 | 4 | 5 |
| Total, all causes | | 8 | 5 | 14 | 19 | 46 | 11 | 10 | 18 | 20 | 59 | 105 |
| Vehicle impacts | I | 1 | 1 | 5 | 6 | 13 | 7 | 7 | 9 | 8 | 31 | 44 |

**Table 6a Further analysis of all significant incidents in terms of use and type of building: 1971–73**

| Type of incident | Code | 2-storey dwellings and unspecified | 1-storey dwellings | Dwellings of 3 storeys and above | Offices | Shops, cafes, public houses | Hospitals, medical | Educational buildings | Other buildings (non-industrial) | Temporary buildings including caravans | Electrical sub-stations | Total all significant incidents 1971–73 (2 years) |
|---|---|---|---|---|---|---|---|---|---|---|---|---|
| Explosions | | | | | | | | | | | | |
| Town gas | T | 23 | 2 | 12 | 1 | 8 | 0 | 0 | 0 | 0 | 0 | 46 |
| Natural gas | N | 22 | 4 | 6 | 0 | 5 | 0 | 1 | 2 | 0 | 1 | 41 |
| Piped LPG | LPG | 0 | 0 | 0 | 0 | 0 | 0 | 0 | 0 | 0 | 0 | 0 |
| Total piped gas | | 45 | 6 | 18 | 1 | 13 | 0 | 1 | 2 | 0 | 1 | 87 |
| Liquefied gas (cylinders, aerosols) | C | 3 | 0 | 0 | 0 | 1 | 0 | 0 | 1 | 0 | 0 | 5 |
| Other materials (dust, vapour, chemicals) | P/S/O | 1 | 1 | 0 | 0 | 2 | 1 | 2 | 1 | 0 | 0 | 8 |
| Water vapour | W | 6 | 0 | 1 | 0 | 0 | 0 | 0 | 0 | 0 | 0 | 7 |
| Lightning | X | 0 | 0 | 0 | 0 | 0 | 0 | 0 | 0 | 0 | 0 | 0 |
| Total other known causes | | 10 | 1 | 1 | 0 | 3 | 1 | 2 | 2 | 0 | 0 | 20 |
| Unknown causes | U | 1 | 0 | 0 | 0 | 0 | 0 | 0 | 0 | 0 | 0 | 1 |
| Total, all causes | | 56 | 7 | 19 | 1 | 16 | 1 | 3 | 4 | 0 | 1 | 108 |
| Vehicle impacts | I | 87 | 9 | 1 | 0 | 15 | 0 | 2 | 5 | 0 | 0 | 119 |

**Table 6b Further analysis of all significant incidents in terms of use and type of building: 1973–75**

| Type of incident | Code | 2-storey dwellings and unspecified | 1-storey dwellings | Dwellings of 3 storeys and above | Offices | Shops, cafes, public houses | Hospitals, medical | Educational buildings | Other buildings (non-industrial) | Temporary buildings including caravans | Electrical sub-stations | Total all significant incidents 1973–75 (2 years) |
|---|---|---|---|---|---|---|---|---|---|---|---|---|
| Explosions | | | | | | | | | | | | |
| Town gas | T | 12 | 0 | 1 | 0 | 0 | 0 | 1 | 0 | 0 | 0 | 14 |
| Natural gas | N | 37 | 4 | 5 | 2 | 5 | 0 | 0 | 3 | 0 | 0 | 56 |
| Piped LPG | LPG | 0 | 0 | 0 | 0 | 0 | 0 | 0 | 0 | 0 | 0 | 0 |
| Total piped gas | | 49 | 4 | 6 | 2 | 5 | 0 | 1 | 3 | 0 | 0 | 70 |
| Liquefied gas (cylinders, aerosols) | C | 7 | 1 | 0 | 0 | 0 | 1 | 0 | 5 | 0 | 0 | 14 |
| Other materials (dust, vapour, chemicals) | P/S/O | 0 | 0 | 0 | 0 | 1 | 0 | 0 | 2 | 0 | 0 | 3 |
| Water vapour | W | 4 | 0 | 1 | 0 | 0 | 0 | 0 | 0 | 0 | 0 | 5 |
| Lightning | X | 1 | 0 | 0 | 0 | 0 | 0 | 0 | 0 | 0 | 0 | 1 |
| Total other known causes | | 12 | 1 | 1 | 0 | 1 | 1 | 0 | 7 | 0 | 0 | 23 |
| Unknown causes | U | 2 | 0 | 0 | 0 | 0 | 0 | 0 | 0 | 0 | 1 | 3 |
| Total, all causes | | 63 | 5 | 7 | 2 | 6 | 1 | 1 | 10 | 0 | 1 | 96 |
| Vehicle impacts | I | 28 | 3 | 0 | 0 | 8 | 0 | 0 | 0 | 0 | 0 | 39 |

**Table 6c Further analysis of all significant incidents in terms of use and type of building: 1975–77**

| Type of incident | Code | 2-storey dwellings and unspecified | 1-storey dwellings | Dwellings of 3 storeys and above | Offices | Shops, cafes, public houses | Hospitals, medical | Educational buildings | Other buildings (non-industrial) | Temporary buildings including caravans | Electrical sub-stations | Total all significant incidents 1975–77 (2 years) |
|---|---|---|---|---|---|---|---|---|---|---|---|---|
| Explosions | | | | | | | | | | | | |
| Town gas | T | 0 | 0 | 1 | 0 | 0 | 1 | 0 | 0 | 0 | 0 | 2 |
| Natural gas | N | 49 | 8 | 13 | 1 | 3 | 0 | 2 | 1 | 0 | 0 | 77 |
| Piped LPG | LPG | 1 | 0 | 0 | 0 | 0 | 0 | 0 | 0 | 0 | 0 | 1 |
| Total piped gas | | 50 | 8 | 14 | 1 | 3 | 1 | 2 | 1 | 0 | 0 | 80 |
| Liquefied gas (cylinders, aerosols) | C | 4 | 2 | 0 | 2 | 2 | 0 | 0 | 4 | 0 | 0 | 14 |
| Other materials (dust, vapour, chemicals) | P/S/O | 0 | 0 | 0 | 0 | 0 | 0 | 0 | 1 | 0 | 0 | 1 |
| Water vapour | W | 3 | 0 | 0 | 0 | 0 | 0 | 0 | 1 | 0 | 0 | 4 |
| Lightning | X | 1 | 0 | 0 | 0 | 0 | 0 | 0 | 0 | 0 | 0 | 1 |
| Total other known causes | | 8 | 2 | 0 | 2 | 2 | 0 | 0 | 6 | 0 | 0 | 20 |
| Unknown causes | U | 0 | 0 | 0 | 0 | 2 | 0 | 0 | 0 | 0 | 3 | 5 |
| Total, all causes | | 58 | 10 | 14 | 3 | 7 | 1 | 2 | 7 | 0 | 3 | 105 |
| Vehicle impacts | I | 31 | 4 | 0 | 0 | 4 | 0 | 0 | 5 | 0 | 0 | 44 |

**Table 6d Further analysis of all significant incidents in terms of use and type of building: 1971–77**

| Type of incident | Code | 2-storey dwellings and unspecified | 1-storey dwellings | Dwellings of 3 storeys and above | Offices | Shops, cafes, public houses | Hospitals, medical | Educational buildings | Other buildings (non-industrial) | Temporary buildings including caravans | Electrical sub-stations | Total all significant incidents 1971–77 (6 years) . |
|---|---|---|---|---|---|---|---|---|---|---|---|---|
| Explosions | | | | | | | | | | | | |
| Town gas | T | 35 | 2 | 14 | 1 | 8 | 1 | 1 | 0 | 0 | 0 | 62 |
| Natural gas | N | 108 | 16 | 24 | 3 | 13 | 0 | 3 | 6 | 0 | 1 | 174 |
| Piped LPG | LPG | 1 | 0 | 0 | 0 | 0 | 0 | 0 | 0 | 0 | 0 | 1 |
| Total piped gas | | 144 | 18 | 38 | 4 | 21 | 1 | 4 | 6 | 0 | 1 | 237 |
| Liquefied gas (cylinders, aerosols) | C | 14 | 3 | 0 | 2 | 3 | 1 | 0 | 10 | 0 | 0 | 33 |
| Other materials (dust, vapour, chemicals) | P/S/O | 1 | 1 | 0 | 0 | 3 | 1 | 2 | 4 | 0 | 0 | 12 |
| Water vapour | W | 13 | 0 | 2 | 0 | 0 | 0 | 0 | 1 | 0 | 0 | 16 |
| Lightning | X | 2 | 0 | 0 | 0 | 0 | 0 | 0 | 0 | 0 | 0 | 2 |
| Total other known causes | | 30 | 4 | 2 | 2 | 6 | 2 | 2 | 15 | 0 | 0 | 63 |
| Unknown causes | U | 3 | 0 | 0 | 0 | 2 | 0 | 0 | 0 | 0 | 4 | 9 |
| Total, all causes | | 177 | 22 | 40 | 6 | 29 | 3 | 6 | 21 | 0 | 5 | 309 |
| Vehicle impacts | I | 146 | 16 | 1 | 0 | 27 | 0 | 2 | 10 | 0 | 0 | 202 |

**Table 7 Further analysis of all significant incidents in terms of severity of incident: annual totals 1971–77**

| Type of incident | Code | 1971–72 | | | 1972–73 | | | 1973–74 | | | 1974–75 | | | 1975–76 | | | 1976–77 | | | Total 1971–77 | | |
|---|---|---|---|---|---|---|---|---|---|---|---|---|---|---|---|---|---|---|---|---|---|---|
| | | M | S | VS | M | S | VS | M | S | VS | M | S | VS | M | S | VS | M | S | VS | M | S | VS |
| Explosions | | | | | | | | | | | | | | | | | | | | | | |
| Town gas | T | 18 | 9 | 4 | 5 | 10 | 0 | 4 | 2 | 0 | 6 | 2 | 0 | 1 | 0 | 0 | 1 | 0 | 0 | 35 | 23 | 4 |
| Natural gas | N | 10 | 10 | 0 | 13 | 4 | 4 | 16 | 8 | 2 | 18 | 9 | 3 | 27 | 7 | 2 | 23 | 17 | 1 | 107 | 55 | 12 |
| Piped LPG | LPG | 0 | 0 | 0 | 0 | 0 | 0 | 0 | 0 | 0 | 0 | 0 | 0 | 0 | 0 | 0 | 0 | 1 | 0 | 0 | 1 | 0 |
| Total piped gas | | 28 | 19 | 4 | 18 | 14 | 4 | 20 | 10 | 2 | 24 | 11 | 3 | 28 | 7 | 2 | 24 | 18 | 1 | 142 | 79 | 16 |
| Liquefied gas (cylinders) | C | 4 | 0 | 0 | 1 | 0 | 0 | 6 | 1 | 0 | 7 | 0 | 0 | 4 | 2 | 0 | 6 | 2 | 0 | 28 | 5 | 0 |
| Other materials (dust, vapour, chemicals) | P/S/O | 3 | 3 | 0 | 1 | 1 | 0 | 0 | 2 | 0 | 1 | 0 | 0 | 0 | 0 | 0 | 0 | 1 | 0 | 5 | 7 | 0 |
| Water vapour | W | 3 | 3 | 0 | 1 | 0 | 0 | 2 | 1 | 0 | 1 | 1 | 0 | 2 | 0 | 0 | 1 | 1 | 0 | 10 | 6 | 0 |
| Lightning | X | 0 | 0 | 0 | 0 | 0 | 0 | 0 | 0 | 0 | 1 | 0 | 0 | 0 | 0 | 0 | 1 | 0 | 0 | 2 | 0 | 0 |
| Total other known causes | | 10 | 6 | 0 | 3 | 1 | 0 | 8 | 4 | 0 | 10 | 1 | 0 | 6 | 2 | 0 | 8 | 4 | 0 | 45 | 18 | 0 |
| Unknown causes | U | 0 | 0 | 0 | 1 | 0 | 0 | 0 | 1 | 0 | 1 | 1 | 0 | 1 | 0 | 0 | 4 | 0 | 0 | 7 | 2 | 0 |
| Total all causes | | 38 | 25 | 4 | 22 | 15 | 4 | 28 | 15 | 2 | 35 | 13 | 3 | 35 | 9 | 2 | 36 | 22 | 1 | 194 | 99 | 16 |
| Vehicle impacts | I | 68 | 0 | 0 | 46 | 5 | 0 | 20 | 0 | 0 | 18 | 1 | 0 | 11 | 2 | 0 | 30 | 1 | 0 | 193 | 9 | 0 |

M = moderate severity; S = severe; VS = very severe

**Table 8a Totals of all significant explosions analysed in terms of use and type of building with corresponding average annual risks: 1971–73**

| Type of building | Code | 1971–73 (2 years) | | | | | |
|---|---|---|---|---|---|---|---|
| | | Total stock of buildings (thousands) 1.4.72* | No of explosions | | No of units affected | | |
| | | | Total† | Risk‡ | Total† | Risk‡ | Average number per incident |
| Detached houses | | 1 964 | 3.3 | 0.8 | 3.3 | 0.8 | 1.0 |
| Semi-detached houses | | 6 180 | 21.1 | 1.7 | 23.3 | 1.9 | 1.1 |
| Terraced houses | | 5 844 | 28.8 | 2.5 | 55.6 | 4.9 | 2.0 |
| | | | (1.0) | | (2.0) | | (2.0) |
| Bungalows | | 1 485 | 7.8 | 2.6 | 7.8 | 2.6 | 1.0 |
| Total, houses & bungalows | D | 15 473 | 61.0 | 2.0 | 90.0 | 3.0 | 1.5 |
| | | | (1.0) | | (2.0) | | (2.0) |
| Separate flats | | 2 884 | 20.0 | 3.5 | 31.0 | 5.4 | 1.6 |
| Flats over shops | | 316 | 1.0 | 1.6 | 1.0 | 12.7 | 1.0 |
| | | | (6.0) | | (1.0) | | (1.2) |
| Total flats | F | 3 200 | 21.0 | 3.3 | 32.0 | 6.1 | 1.5 |
| | | | (6.0) | | (7.0) | | (1.2) |
| Total, dwellings | | 18 673 | 82.0 | 2.2 | 122.0 | 3.5 | 1.5 |
| | | | (7.0) | | (9.0) | | (1.3) |
| Shops, cafes, public houses, restaurants, stores, garages and warehouses | S/W | 1 086 | 16.0 | 7.4 | 31.0 | 14.3 | 1.9 |
| Hotels, hostels, public conveniences, laboratories, cultural, educational, medical, entertainment, religious, offices | H/I/X/O | 500 | 7.0 | 7.0 | 7.0 | 7.0 | 1.0 |
| Electrical or gas sub-stations, telephone exchanges | E/T | 100 | 3.0 | 15.0 | 3.0 | 15.0 | 1.0 |
| Total,other non-industrial buildings | | 1 686 | 26.0 | 7.7 | 41.0 | 12.2 | 1.6 |

*The total stocks of buildings have been derived at by reference to figures published by the Inland Revenue, the DOE, and Audits of Great Britain Ltd. Figures for the last two non-dwelling groups are more approximate estimates to give approximate risk figures

†Explosions in dwellings of unknown type have been distributed between houses and bungalows in proportion to the known numbers of explosions in these types

‡Risk is expressed in terms of units per million units per year

Figures in parentheses relate to explosions resulting in damage to dwellings when the explosion itself was in a shop in the same building

**Table 8b Totals of all significant explosions analysed in terms of use and type of building with corresponding average annual risks: 1973–75**

| Type of building | Code | 1973–75 (2 years) | | | | | |
|---|---|---|---|---|---|---|---|
| | | Total stock of buildings (thousands) 1.4.74* | No of explosions | | No of units affected | | |
| | | | Total† | Risk‡ | Total† | Risk‡ | Average number per incident |
| Detached houses | | 1 962 | 9.0 | 2.3 | 9.0 | 2.3 | 1.0 |
| Semi-detached houses | | 6 252 | 27.5<br>(1.0) | 2.2 | 39.5<br>(2.0) | 3.3 | 1.4<br>(2.0) |
| Terraced houses | | 5 850 | 22.8 | 1.9 | 32.8 | 2.8 | 1.4 |
| Bungalows | | 1 497 | 5.7 | 1.9 | 5.7 | 1.9 | 1.0 |
| Total, houses & bungalows | D | 15 561 | 65.0<br>(1.0) | 2.1 | 87.0<br>(2.0) | 2.9 | 1.3<br>(2.0) |
| Separate flats | | 3 237 | 9.0 | 1.4 | 16.0 | 2.5 | 1.8 |
| Flats over shops | | 295 | 1.0<br>(4.0) | 1.7 | 1.0<br>(9.0) | 16.9 | 1.0<br>(2.3) |
| Total flats | F | 3 532 | 10.0<br>(4.0) | 1.4 | 17.0<br>(9.0) | 3.7 | 1.7<br>(2.3) |
| Total, dwellings | | 19 093 | 75.0<br>(5.0) | 2.0 | 104.0<br>(11.0) | 3.0 | 1.4<br>(2.2) |
| Shops, cafes, public houses, restaurants, stores, garages and warehouses | S/W | 1 115 | 11.0 | 4.9 | 21.0 | 9.4 | 1.9 |
| Hotels, hostels, public conveniences, laboratories, cultural, educational, medical, entertainment, religious, offices | H/I/X/O | 500 | 7.0 | 7.0 | 7.0 | 7.0 | 1.0 |
| Electrical or gas sub-stations, telephone exchanges | E/T | 100 | 3.0 | 15.0 | 3.0 | 15.0 | 1.0 |
| Total,other non-industrial buildings | | 1 715 | 21.0 | 6.1 | 31.0 | 9.0 | 1.5 |

*The total stocks of buildings have been derived at by reference to figures published by the Inland Revenue, the DOE, and Audits of Great Britain Ltd. Figures for the last two non-dwelling groups are more approximate estimates to give approximate risk figures

†Explosions in dwellings of unknown type have been distributed between houses and bungalows in proportion to the known numbers of explosions in these types

‡Risk is expressed in terms of units per million units per year

Figures in parentheses relate to explosions resulting in damage to dwellings when the explosion itself was in a shop in the same building

**Table 8c Totals of all significant explosions analysed in terms of use and type of building with corresponding average annual risks: 1975–77**

| Type of building | Code | 1975–77 (2 years) Total stock of buildings (thousands) 1.4.76 (estimated)* | No of explosions Total† | No of explosions Risk‡ | No of units affected Total† | No of units affected Risk‡ | Average number per incident |
|---|---|---|---|---|---|---|---|
| Detached houses | | 1 960 | 5.4 | 1.4 | 5.4 | 1.4 | 1.0 |
| Semi-detached houses | | 6 546 | 24.2 | 1.8 | 28.8 | 2.2 | 1.2 |
| Terraced houses | | 5 860 | 24.6 | 2.1 | 31.0 | 2.6 | 1.3 |
| Bungalows | | 1 510 | 9.8 | 3.2 | 9.8 | 3.2 | 1.0 |
| Total, houses & bungalows | D | 15 876 | 64.0 | 2.0 | 75.0 | 2.4 | 1.2 |
| Separate flats | | 3 444 | 16.0 | 2.3 | 37.0 | 5.4 | 2.3 |
| Flats over shops | | 280 | 2.0 (1.0) | 3.6 | 4.0 (2.0) } | 10.7 | 2.0 (2.0) |
| Total flats | F | 3 724 | 18.0 (1.0) | 2.4 | 41.0 (2.0) } | 5.8 | 2.3 (2.0) |
| Total, dwellings | | 19 600 | 82.0 (1.0) | 2.1 | 116.0 (2.0) } | 3.0 | 1.4 (2.0) |
| Shops, cafes, public houses, restaurants, stores, garages and warehouses | S/W | 1 110 | 11.0 | 5.0 | 21.0 | 9.5 | 1.9 |
| Hotels, hostels, public conveniences, laboratories, cultural, educational, medical, entertainment, religious, offices | H/I/X/O | 500 | 8.0 | 8.0 | 10.0 | 10.0 | 1.3 |
| Electrical or gas sub-stations, telephone exchanges | E/T | 100 | 4.0 | 20.0 | 4.0 | 20.0 | 1.0 |
| Total,other non-industrial buildings | | 1 710 | 23.0 | 6.7 | 35.0 | 10.2 | 1.5 |

*The total stocks of buildings have been derived at by reference to figures published by the Inland Revenue, the DOE, and Audits of Great Britain Ltd. Figures for the last two non-dwelling groups are more approximate estimates to give approximate risk figures

†Explosions in dwellings of unknown type have been distributed between houses and bungalows in proportion to the known numbers of explosions in these types

‡Risk is expressed in terms of units per million units per year

Figures in parentheses relate to explosions resulting in damage to dwellings when the explosion itself was in a shop in the same building

**Table 8d Totals of all significant explosions analysed in terms of use and type of building with corresponding average annual risks: 1971–77**

| Type of building | Code | Total stock of buildings (thousands) 1.4.74* | No of explosions: Total† | No of explosions: Risk‡ | No of units affected: Total† | No of units affected: Risk‡ | No of units affected: Average number per incident |
|---|---|---|---|---|---|---|---|
| | | | 1971–77 (6 years) | | | | |
| Detached houses | | 1 962 | 17.7 | 1.5 | 17.7 | 1.5 | 1.0 |
| Semi-detached houses | | 6 252 | 72.8<br>(1.0) | 1.9 | 91.6<br>(2.0) | 2.5 | 1.3<br>(2.0) |
| Terraced houses | | 5 850 | 76.2<br>(1.0) | 2.2 | 119.4<br>(2.0) | 3.5 | 1.6<br>(2.0) |
| Bungalows | | 1 497 | 23.3 | 2.6 | 23.3 | 2.6 | 1.0 |
| Total, houses & bungalows | D | 15 561 | 190.0<br>(2.0) | 2.0 | 252.0<br>(4.0) | 2.7 | 1.3<br>(2.0) |
| Separate flats | | 3 237 | 45.0 | 2.3 | 84.0 | 4.3 | 1.9 |
| Flats over shops | | 295 | 4.0<br>(11.0) | 2.3 | 6.0<br>(18.0) | 13.6 | 1.5<br>(1.6) |
| Total flats | F | 3 532 | 49.0<br>(11.0) | 2.3 | 90.0<br>(18.0) | 5.1 | 1.8<br>(1.6) |
| Total, dwellings | | 19 093 | 239.0<br>(13.0) | 2.1 | 342.0<br>(22.0) | 3.2 | 1.4<br>(1.7) |
| Shops, cafes, public houses, restaurants, stores, garages and warehouses | S/W | 1 115 | 38.0 | 5.7 | 73.0 | 10.9 | 1.9 |
| Hotels, hostels, public conveniences, laboratories, cultural, educational, medical, entertainment, religious, offices | H/I/X/O | 500 | 22.0 | 7.3 | 24.0 | 8.0 | 1.1 |
| Electrical or gas sub-stations, telephone exchanges | E/T | 100 | 10.0 | 16.7 | 10.0 | 16.7 | 1.0 |
| Total,other non-industrial buildings | | 1 715 | 70.0 | 6.8 | 107.0 | 10.4 | 1.5 |

*The total stocks of buildings have been derived at by reference to figures published by the Inland Revenue, the DOE, and Audits of Great Britain Ltd. Figures for the last two non-dwelling groups are more approximate estimates to give approximate risk figures

†Explosions in dwellings of unknown type have been distributed between houses and bungalows in proportion to the known numbers of explosions in these types

‡Risk is expressed in terms of units per million units per year

Figures in parentheses relate to explosions resulting in damage to dwellings when the explosion itself was in a shop in the same building

**Table 9 Totals of all significant explosions in dwellings analysed in terms of age of building: 1971–73; 1973–75; 1975–77; 1971–77**

| Date of construction | Age of dwelling at time of incident (years) | 1971–73 (2 years) | | | 1973–75 (2 years) | | | 1975–77 (2 years) | | | 1971–77 (6 years) | | |
|---|---|---|---|---|---|---|---|---|---|---|---|---|---|
| | | % of stock* | No of incidents | % of total | % of stock † | No of incidents | % of total | % of stock‡ | No of incidents | % of total | % of stock § | No of incidents | % of total |
| Pre-1919 | 52 or more | 35 | 33 | 40 | 34 | 29 | 39 | 33 | 38 | 46 | 34 | 100 | 42 |
| 1919–45 | 26 to 58 | 24 | 17 | 21 | 23 | 19 | 25 | 22 | 13 | 16 | 23 | 49 | 20 |
| Post-1945 | 32 or less | 41 | 32 | 39 | 43 | 27 | 36 | 45 | 31 | 38 | 43 | 90 | 38 |
| Totals | | 100 | 82 | 100 | 100 | 75 | 100 | 100 | 82 | 100 | 100 | 239 | 100 |

Percentages of total dwelling stock are estimated as follows: *December 1971, †December 1973, ‡December 1975, §December 1973

**Table 10 Totals of all significant explosions in dwellings analysed in terms of height of building: 1971–73; 1973–75; 1975–77; 1971–77**

| No of storeys | Estimated % of total dwelling stock | 1971–73 (2 years) | | | | 1973–75 (2 years) | | | | 1975–77 (2 years) | | | | 1971–77 (6 years) | | | |
|---|---|---|---|---|---|---|---|---|---|---|---|---|---|---|---|---|---|
| | | No of incidents | % of total | No of dwellings affected | % of total | No of incidents | % of total | No of dwellings affected | % of total | No of incidents | % of total | No of dwellings affected | % of total | No of incidents | % of total | No of dwellings affected | % of total |
| 1 (bungalows) | 8 | 8 | 10 | 8 | 6 | 6 | 8 | 6 | 5 | 10 | 12 | 10 | 8½ | 24 | 10 | 24 | 6½ |
| 2<br>3 | 87 | 53<br>14 | 82 | 83<br>26 | 83 | 61<br>6 | 89⅓ | 89<br>15 | 90½ | 57<br>9 | 81 | 72<br>26 | 83 | 171<br>29 | 83½ | 244<br>67 | 85½ |
| 4 | 2 | 4 | 5 | 7 | 5 | 0 | 0 | 0 | 0 | 2 | 2½ | 2 | 2 | 6 | 2½ | 9 | 2½ |
| 5–9 | 1 | 1 | 1 | 5 | 4 | 1 | 1⅓ | 4 | 3½ | 2 | 2½ | 3 | 2½ | 4 | 2 | 12 | 3 |
| 10–14 | 1 | 0 | 0 | 0 | 0 | 1 | 1⅓ | 1 | 1 | 1 | 1 | 1 | 1 | 2 | 1 | 2 | ½ |
| 15 or more | 1 | 2 | 2 | 2 | 2 | 0 | 0 | 0 | 0 | 1 | 1 | 4 | 3 | 3 | 1 | 6 | 2 |
| Total | 100 | 82 | 100 | 131 | 100 | 75 | 100 | 115 | 100 | 82 | 100 | 118 | 100 | 239 | 100 | 364 | 100 |

**Table 11 Annual totals of significant explosions of piped gas and liquefied gas in relation to gas consumption and numbers of consumers**

| | 1971–72 | 1972–73 | 1973–74 | 1974–75 | 1975–76 | 1976–77 | 1971–77 |
|---|---|---|---|---|---|---|---|
| Town gas | | | | | | | |
| Significant incidents | 31 | 15 | 6 | 8 | 1 | 1 | 62 |
| Consumption/$10^9$ MJ† | 229 | 161 | 114 | 67 | 26 | 7 | 604 |
| Risk/$10^9$ MJ | 0.135 | 0.093 | 0.053 | 0.119 | 0.038 | 0.143 | 0.103 |
| No of consumers/$10^6$ † | 8.15 | 6.05 | 4.10 | 2.36 | 1.20 | 0.55 | (22.41) |
| Risk/$10^6$ consumers‡ | 3.8 | 2.5 | 1.5 | 3.4 | 0.8 | 1.8 | 2.8 |
| Natural gas | | | | | | | |
| Significant incidents | 20 | 21 | 26 | 30 | 36 | 41 | 174 |
| Consumption/$10^9$ MJ† | 291 | 435 | 539 | 673 | 753 | 808 | 3499 |
| Risk/$10^9$ MJ | 0.069 | 0.048 | 0.048 | 0.045 | 0.048 | 0.051 | 0.050 |
| No of consumers/$10^6$ † | 4.63 | 6.81 | 8.86 | 10.69 | 12.04 | 12.96 | (55.99) |
| Risk/$10^6$ consumers‡ | 4.3 | 3.1 | 2.9 | 2.8 | 3.0 | 3.2 | 3.1 |
| All piped gas* | | | | | | | |
| Significant incidents | 51 | 36 | 32 | 38 | 37 | 42* | 236* |
| Consumption/$10^9$ MJ† | 520 | 596 | 653 | 740 | 779 | 815 | 4103 |
| Risk/$10^9$ MJ | 0.098 | 0.060 | 0.048 | 0.051 | 0.047 | 0.052 | 0.058 |
| No of consumers/$10^6$ † | 12.78 | 12.86 | 12.96 | 13.05 | 13.24 | 13.51 | (78.40) |
| Risk/$10^6$ consumers‡ | 4.0 | 2.8 | 2.5 | 2.9 | 2.8 | 3.1 | 3.0 |
| Cylinders | | | | | | | |
| Significant incidents | 4 | 1 | 7 | 7 | 6 | 8 | 33 |
| Consumption/$10^9$ MJ† | 5.4 | 5.8 | 6.3 | 6.3 | 6.3 | 6.3 | 36.4 |
| Risk/$10^9$ MJ | 0.74 | 0.17 | 1.11 | 1.11 | 0.95 | 1.27 | 0.9 |

*The single incident that involved piped liquefied petroleum gas is not included in this table

†Figures for consumption and numbers of consumers have been derived from British Gas data and the Digest of United Kingdom Energy Statistics

‡Risks per million consumers are annual risks

**Table 12 Annual totals of significant explosions of piped gas and liquefied gas broken down in terms of type and source of gas and severity of incident**

| | | 1971–72 | | | 1972–73 | | | 1973–74 | | | 1974–75 | | | 1975–76 | | | 1976–77 | | | 1971–77 | | |
|---|---|---|---|---|---|---|---|---|---|---|---|---|---|---|---|---|---|---|---|---|---|---|
| Type of explosions | Code | M | S | VS | M | S | VS | M | S | VS | M | S | VS | M | S | VS | M | S | VS | M | S | VS |
| Town gas | | | | | | | | | | | | | | | | | | | | | | |
| External/service pipe | E | 4 | 3 | 1 | 0 | 2 | 0 | 2 | 0 | 0 | 1 | 0 | 0 | 0 | 0 | 0 | 0 | 0 | 0 | 7 | 5 | 1 |
| Meter/installation pipe | I | 5 | 1 | 0 | 0 | 4 | 0 | 0 | 1 | 0 | 3 | 1 | 0 | 1 | 0 | 0 | 1 | 0 | 0 | 10 | 7 | 0 |
| Appliance | A | 7 | 4 | 3 | 3 | 2 | 0 | 0 | 1 | 0 | 1 | 1 | 0 | 0 | 0 | 0 | 0 | 0 | 0 | 11 | 8 | 3 |
| Unknown | U | 2 | 1 | 0 | 2 | 2 | 0 | 2 | 0 | 0 | 1 | 0 | 0 | 0 | 0 | 0 | 0 | 0 | 0 | 7 | 3 | 0 |
| Total | | 18 | 9 | 4 | 5 | 10 | 0 | 4 | 2 | 0 | 6 | 2 | 0 | 1 | 0 | 0 | 1 | 0 | 0 | 35 | 23 | 4 |
| Natural gas | | | | | | | | | | | | | | | | | | | | | | |
| External/service pipe | E | 2 | 7 | 0 | 5 | 1 | 2 | 8 | 5 | 2 | 8 | 6 | 2 | 9 | 1 | 1 | 7 | 9 | 1 | 39 | 29 | 8 |
| Meter/installation pipe | I | 6 | 3 | 0 | 2 | 1 | 1 | 1 | 3 | 0 | 5 | 3 | 1 | 9 | 5 | 1 | 13 | 8 | 0 | 36 | 23 | 3 |
| Appliance | A | 1 | 0 | 0 | 3 | 2 | 0 | 5 | 0 | 0 | 4 | 0 | 0 | 5 | 1 | 0 | 2 | 0 | 0 | 20 | 3 | 0 |
| Unknown | U | 1 | 0 | 0 | 3 | 0 | 1 | 2 | 0 | 0 | 1 | 0 | 0 | 4 | 0 | 0 | 1 | 0 | 0 | 12 | 0 | 1 |
| Total | | 10 | 10 | 0 | 13 | 4 | 4 | 16 | 8 | 2 | 18 | 9 | 3 | 27 | 7 | 2 | 23 | 17 | 1 | 107 | 55 | 12 |
| All piped gas* | | | | | | | | | | | | | | | | | | | | | | |
| External/service pipe | E | 6 | 10 | 1 | 5 | 3 | 2 | 10 | 5 | 2 | 9 | 6 | 2 | 9 | 1 | 1 | 7 | 9 | 1 | 46 | 34 | 9 |
| Meter/installation pipe | I | 11 | 4 | 0 | 2 | 5 | 1 | 1 | 4 | 0 | 8 | 4 | 1 | 10 | 5 | 1 | 14 | 8 | 0 | 46 | 30 | 3 |
| Appliance | A | 8 | 4 | 3 | 6 | 4 | 0 | 5 | 1 | 0 | 5 | 1 | 0 | 5 | 1 | 0 | 2 | 0 | 0 | 31 | 11 | 3 |
| Unknown | U | 3 | 1 | 0 | 5 | 2 | 1 | 4 | 0 | 0 | 2 | 0 | 0 | 4 | 0 | 0 | 1 | 0 | 0 | 19 | 3 | 1 |
| Total | | 28 | 19 | 4 | 18 | 14 | 4 | 20 | 10 | 2 | 24 | 11 | 3 | 28 | 7 | 2 | 24 | 17 | 1 | 142 | 78 | 16 |
| LPG | | | | | | | | | | | | | | | | | | | | | | |
| Cylinders | | 4 | 0 | 0 | 1 | 0 | 0 | 6 | 1 | 0 | 7 | 0 | 0 | 4 | 2 | 0 | 6 | 2 | 0 | 28 | 5 | 0 |

*The single incident that involved piped liquefied petroleum gas is not included in this table
M = moderate severity; S = severe; VS = very severe

**Table 13 Some major accidental explosions abroad during the period of the present survey**

| Date | Place | Building(s) affected | Estimated cost* | Fatalities† |
|---|---|---|---|---|
| 1.12.71 | Naples, Italy | 3-storey office + flats | A | 9 |
| 21.12.71 | Argenteuil, France | 14-storey flats | B | 20 |
| 6.3.72 | Barcelona, Spain | 10-storey flats | A | 18 |
| 12.6.72 | Munich, Germany | 5-storey flats | A | 4 |
| 30.10.72 | Barcelona, Spain | ?-storey flats (3 blocks) | A | 14 |
| 30.11.72 | Rome, Italy‡ | 9-storey flats + shops | A | 16 |
| 1.6.73 | Perpignan, France | 5-storey flats | A | 7 |
| 2.12.73 | Copenhagen, Denmark | 5-storey flats | A | 2 |
| 13.12.73 | Tachov, Czechoslovakia | 4-storey workers' hostel | A | 50 |
| 22.4.74 | New York, USA | 24-storey offices | B | 0 |
| 10.1.75 | Freemont,Nebraska, USA | 5-storey hotel | A | 18 |
| 16.1.76 | Tiaret, Algeria | 3-storey hotel/3-storey flats | A | 35 |
| 2.11.76 | Szolnok, Hungary | 4-storey flats | A | 12 |

*Rough estimates only at 1977 UK prices: A = £100 000 to £500 000, B = over £500 000
†Mostly reported initial confirmed figures – full totals may have been higher
‡This incident may not have been caused by piped gas. Most or all of the others were

**Table 14 Totals of all significant vehicle impacts analysed in terms of use and type of building with corresponding average annual risks: 1971–73; 1973–75; 1975–77; 1971–77***

| Type of building | Code | 1971–73 (2 years) | | 1973–75 (2 years) | | 1975–77 (2 years) | | 1971–77 (6 years) | |
|---|---|---|---|---|---|---|---|---|---|
| | | No of impacts† | Risk‡ | No of impacts† | Risk‡ | No of impacts† | Risk‡ | No of impacts† | Risk‡ |
| Detached houses | | 5.9 | 1.5 | 3.5 | 0.9 | 4.8 | 1.2 | 14.2 | 1.2 |
| Semi-detached houses | | 26.4 | 2.1 | 12.2 | 1.0 | 17.4 | 1.3 | 56.0 | 1.5 |
| Terraced houses | | 46.1 | 3.9 | 12.2 | 1.0 | 6.4 | 0.5 | 64.7 | 1.9 |
| Bungalows | | 17.6 | 5.9 | 5.1 | 1.7 | 6.4 | 2.1 | 29.1 | 3.2 |
| Total houses & bungalows | D | 96 | 3.1 | 33 | 1.1 | 35 | 1.1 | 164 | 1.8 |
| Separate flats | | 2 | 0.3 | 2 | 0.3 | 0 | 0 | 4 | 0.2 |
| Flats over shops | | 4 | 6.3 | 0 | 0 | 1 | 1.8 | 5 | 2.8 |
| Total flats | F | 6 | 0.9 | 2 | 0.3 | 1 | 0.1 | 9 | 0.4 |
| Total dwellings | | 102 | 2.7 | 35 | 0.9 | 36 | 0.9 | 173 | 1.5 |
| Shops, public houses, garages etc | S/W | 20 | 9.2 | 9 | 4.0 | 7 | 3.2 | 36 | 5.4 |
| Hotels, hostels, educational etc | H/I/X | 4 | 4.0 | 0 | 0 | 3 | 3.0 | 7 | 2.3 |
| Sub-stations | E | 0 | 0 | 0 | 0 | 1 | 5.0 | 1 | 1.7 |
| Total other non-industrial buildings | | 24 | 7.1 | 9 | 2.6 | 11 | 3.2 | 44 | 4.3 |

*See Section 1.6 for a note on the validity of these data

†Incidents in dwellings of unknown type have been distributed between houses and bungalows in proportion to the known numbers of impacts with these types. Numbers given are for the numbers of units hit and significantly damaged

‡Risk is expressed in terms of units per million units per year. For total stocks of buildings see Tables 8a–d

**Table 15 Totals of significant vehicle impacts with dwellings analysed in terms of age of building: 1971–73; 1973–75; 1975–77; 1971–77**

| Date of construction | Age of dwelling at time of impact (years) | 1971–73 (2 years) | | | 1973–75 (2 years) | | | 1975–77 (2 years) | | | 1971–77 (6 years) | | |
|---|---|---|---|---|---|---|---|---|---|---|---|---|---|
| | | % of stock* | No of impacts | % of total | % of stock† | No of impacts | % of total | % of stock‡ | No of impacts | % of total | % of stock § | No of impacts | % of total |
| Pre-1919 | 52 or more | 35 | 68 | 66 | 34 | 19 | 53 | 33 | 28 | 78 | 34 | 115 | 67 |
| 1919–45 | 26 to 58 | 24 | 20 | 20 | 23 | 8 | 24 | 22 | 2 | 6 | 23 | 30 | 17 |
| Post-1945 | 32 or less | 41 | 14 | 14 | 43 | 8 | 23 | 45 | 6 | 16 | 43 | 28 | 16 |
| Totals | | 100 | 102 | 100 | 100 | 35 | 100 | 100 | 36 | 100 | 100 | 173 | 100 |

Percentages of total dwelling stock are estimated as follows: *December 1971, †December 1973, ‡December 1975, § December 1973

**Table 16 Totals of significant vehicle impacts with dwellings analysed in terms of height of building: 1971–73; 1973–75; 1975–77; 1971–77**

| No of storeys | Estimated % of total dwelling stock | 1971–73 (2 years) | | 1973–75 (2 years) | | 1975–77 (2 years) | | 1971–77 (6 years) | |
|---|---|---|---|---|---|---|---|---|---|
| | | No of impacts | % of total | No of impacts | % of total | No of impacts | % of total | No of impacts | % of total |
| 1 (bungalows) | 8 | 18 | 18 | 5 | 14 | 6 | 17 | 29 | 17 |
| 2 and 3 | 87 | 84 | 82 | 30 | 86 | 30 | 83 | 144 | 83 |
| 4 | 2 | 0 | 0 | 0 | 0 | 0 | 0 | 0 | 0 |
| 5 and above | 3 | 0 | 0 | 0 | 0 | 0 | 0 | 0 | 0 |
| Total | 100 | 102 | 100 | 35 | 100 | 36 | 100 | 173 | 100 |

## APPENDICES

### Key

U indicates 'unknown', a dash (–) 'not applicable' or 'no estimate'

| | |
|---|---|
| Use of building: | D – house (including bungalow) |
| | F – flat, maisonette |
| | S – shop, cafe, restaurant, public house |
| | S/(F) etc – dual use of building or buildings affected with use in which explosion occurred shown first |
| | W – store, garage, warehouse |
| | H – hotel, hostel |
| | I – university, school, hospital, clinic |
| | O – office |
| | X – building used by public for religion, culture, entertainment and other purposes |
| | E – sub-station, pumping station |
| | T – telephone exchange |
| Type of building: (house only) | D – detached |
| | S – semi-detached |
| | T – terraced |
| | B – bungalow |
| Age of building: | a – constructed before 1919 |
| | b – constructed 1919–45 |
| | c – constructed after 1945 |
| Structure of building: | B – brickwork |
| | C – concrete |
| | CF – in-situ concrete frame |
| | CP – precast concrete panel |
| | CW – in-situ concrete wall |
| | M – masonry |
| | S – steel |
| | T – timber or timber frame |
| Explosive substance: | T – town gas |
| | N – natural gas |
| | L – piped liquefied petroleum gas |
| | C – cylinder gas, aerosols |
| | P – petroleum vapour |
| | S – sewer gas |
| | O – other gases |
| | W – water vapour |
| | X – lightning |
| Source: (piped gas only) | E – external, service pipe |
| | I – internal, installation pipe |
| | A – appliance |
| Ignition: | F – fire before explosion |
| | L – naked flame |
| | E – electrical |
| | A – appliance (including pilot light) |

## Appendix A Details of all significant explosions

| Incident no | No of fatal casualties | Building | | | | | Explosion | | | Damage (severity) | | | Estimated cost of repairs (£) | Estimated peak pressure (kN/m²) |
|---|---|---|---|---|---|---|---|---|---|---|---|---|---|---|
| | | Use | Type (house only) | No of storeys | Age | Structure | Explosive substance | Source (piped gas only) | Ignition | Extent of significant damage | Fire† | Rating‡ (explosion only) | | |
| **April–June 1971** | | | | | | | | | | | | | | |
| 13 | 0 | D | T | 2 | a | M | T | E | U | 3 houses | Nil | S | 1 000 | – |
| 17 | 0 | S/(F) | – | 2 | a | B | T | I | F | 1 cafe/1 flat | B | S | 7 500 | 10–25 |
| 41 | 0 | S | – | 2 | a | T | T | I | U | 1 restaurant | A | M | 500 | 5–10 |
| 43 | 0 | D | S | 2 | c | B | T | A | L | 1 house | Nil | M | 5 000 | 10–15 |
| 44 | 0 | D | B | 1 | c | B | T | U | U | 1 bungalow | Nil | M | 500 | – |
| 51 | 0 | F | – | 3 | c | U | N | I | L | 1 flat | Nil | M | 500 | – |
| 64 | 0 | D | T | 2 | c | B | N | E | L | 1 house | B | M | 500 | 5–10 |
| 85A | 0 | F | – | 3 | c | B | T | A | F | 3 flats | Nil | VS | 10 000 | 15–30 |
| 103 | 0 | I | – | 2 | c | B | O | – | L | 1 laboratory | A | M | 1 500 | 5–10 |
| 128 | 0 | D | S | 2 | a | M | N | I | L | 1 house | A | M | 1 500 | 5–15 |
| **July–Sept 1971** | | | | | | | | | | | | | | |
| 163 | 0 | S/(F) | – | U | U | U | C | – | U | 1 shop/1 flat | Nil | M | 500 | – |
| 164E | 0 | O | – | 10 | c | CF | T | I | E | 1 basement room | Nil | M | 300 | 10–20 |
| 85B | 0 | F | – | 4 | c | B | T | A | A | 1 flat | Nil | M | 500 | 2.5–7.5 |
| 186 | 0 | I | – | 4 | c | CF | O | – | U | 1 room | Nil | S | 1 000 | – |
| 197 | 0 | S/(F) | – | 2 | b | B | T | A | E | 1 shop/1 flat | Nil | M | 800 | 5–15 |
| 202 | 0 | F | – | 4 | a | B | W | – | – | 1 flat | Nil | S | 1 000 | – |
| 230 | 0 | D | T | 2 | a | B | W | – | – | 1 house | Nil | M | 500 | – |
| 245 | 0 | D | S | 2 | b | B | T | U | U | 1 house | A | M | 1 000 | – |
| 259 | 0 | S/(F) | – | 2 | b | B | N | I | L | 1 cafe | Nil | M | 1 000 | 5–10 |
| 284 | 0 | D | U | U | U | U | C | – | F | 1 house | A | M | 750 | – |
| 304 | 0 | D | S | 2 | c | B | T | I | L | 1 house | Nil | M | 750 | – |
| 330 | 1 | F | – | 4 | b | B | T | A | E | 3 flats | Nil | VS | 20 000 | 25–45 |
| 346 | 0 | D | B | 1 | b | B | N | E | A | 1 bungalow | A | S | 5 000 | 5–15 |
| 361 | 0 | D | D | 2 | a | B | C | – | L | 1 house | Nil | M | 1 000 | 5–15 |
| 382 | 0 | F | – | 2 | a | B | N | I | U | 1 flat | A | S | 1 200 | 15–25 |
| 364 | 0 | S/(F) | – | 3 | a | B/M | N | I | U | 1 public house | Nil | M | 1 200 | 5–10 |
| 365 | 0 | F | – | 6 | c | CF | T | E | U | 4 flats | Nil | M | 1 200 | 10–15 |
| 376 | 0 | S/(F) | – | 3 | a | B/M | N | E | U | 1 shop/1 flat | Nil | M | 6 000 | 5–15 |

†B before; A after.

‡M moderate; S severe; VS very severe.

| Incident no | No of fatal casualties | Building | | | | | Explosion | | | Damage (severity) | | | Estimated cost of repairs (£) | Estimated peak pressure (kN/m$^2$) |
|---|---|---|---|---|---|---|---|---|---|---|---|---|---|---|
| | | Use | Type (house only) | No of storeys | Age | Structure | Explosive substance | Source (piped gas only) | Ignition | Extent of significant damage | Fire† | Rating‡ (explosion only) | | |
| Oct–Dec 1971 | | | | | | | | | | | | | | |
| 375 | 0 | D | T | 2 | c | C | T | A | F | 1 house | A | M | 750 | – |
| 394 | 0 | S | – | 3 | a | U | S | – | L | 1 restaurant | Nil | S | 1 000 | – |
| 409 | 0 | D | S | 2 | c | B | T | E | E | 1 house | Nil | S | 6 000 | 10–15 |
| 431 | 22 | S | – | 2 | c | CF | T | E | U | 13 shops | Nil | VS | 120 000 | 250–500 |
| 442 | 1 | D | S | 2 | a | B | T | A | U | 1 house | Nil | S | 5 000 | 10–20 |
| 457 | 0 | D | T | 3 | a | B | N | E | A | 2 houses | Nil | S | 12 000 | 10–20 |
| 546 | 2 | D | T | 2 | a | B | P | – | E | 1 house | Nil | M | 200 | – |
| 571 | 0 | D | S | 2 | a | B | T | E | F | 2 houses | Nil | S | 10 000 | – |
| 576 | 0 | F | – | 3 | c | B | T | I | A | 1 flat | Nil | M | 400 | – |
| 614 | 0 | D | T | U | U | B | W | – | U | 1 house | Nil | S | 2 000 | – |
| 623 | 0 | D | S | U | c | U | N | I | U | 1 house | A | M | 800 | – |
| 624 | 0 | S/(F) | – | 2 | a | B | T | E | L | 1 public house/1 flat | A | M | 2 500 | 10–15 |
| Jan–Mar 1972 | | | | | | | | | | | | | | |
| 663 | 0 | D | T | 2 | a | B | T | A | A | 1 house | Nil | M | 750 | – |
| 666 | 0 | D | S | 2 | a | M | N | E | F | 1 house | Nil | S | 2 500 | 10–20 |
| 680 | 0 | D | T | 2 | b | B | N | E | L | 1 house | Nil | S | 1 500 | 5–15 |
| 685 | 0 | F | – | 3 | c | B | T | A | U | 3 flats | Nil | S | 3 000 | 10–15 |
| 720 | 1 | E | – | 1 | U | B | N | E | E | 1 sub-station | Nil | S | 2 500 | 10–20 |
| 727 | 0 | H | – | 4 | a | B/M | N | E | E | 3 rooms | A | S | 2 000 | – |
| 758 | 0 | D | S | 2 | c | B | T | I | U | 1 house | Nil | M | 500 | – |
| 766 | 1 | F | – | 3 | c | B | T | A | A | 3 flats | Nil | VS | 20 000 | 15–30 |
| 778 | 0 | D | U | U | U | U | W | – | – | 1 house | Nil | M | 500 | – |
| 779 | 0 | D | T | 3 | a | B | T | U | U | 2 houses | A | S | 6 000 | 5–15 |
| 786 | 0 | D | S | 2 | c | C | N | I | U | 2 houses | A | S | 9 000 | 20–30 |
| 801 | 0 | H | – | 2 | a | B | N | I | L | 2 rooms | Nil | S | 2 000 | 5–15 |
| 803 | 0 | F | – | U | U | U | W | – | – | 1 flat | Nil | M | 500 | – |
| 808 | 0 | D | T | 2 | b | B | N | E | U | 1 house | A | S | 3 000 | – |
| 842 | 0 | D | B | 1 | b | B | T | A | U | 1 bungalow | Nil | M | 1 500 | 5–10 |
| 855 | 0 | D | S | 2 | a | M | N | I | E | 1 house | U | M | 2 500 | – |
| 874 | 0 | D | S | 2 | b | B | W | – | – | 1 house | Nil | S | 5 000 | (10–25)* |
| 902 | 0 | F | – | 16 | c | CW | N | A | L | 1 flat | Nil | M | 500 | 5–10 |
| 923 | 0 | F | – | 17 | c | CF | T | A | U | 1 flat | Nil | M | 1 000 | 5–15 |
| 987 | 0 | D | B | 1 | b | M | O | – | A | 1 bungalow | A | M | 500 | – |
| 997 | 0 | S/(O) | – | 2+ | U | U | N | U | U | 1 cafe | A | M | 2 000 | – |
| 1007 | 0 | D | D | 2 | b | U | T | E | U | 1 house | Nil | M | 500 | – |
| 1009 | 0 | E | – | 1 | c | B | C | – | U | 1 pumping station | Nil | M | 500 | – |

*Water heater

| Incident no | No of fatal casualties | Building | | | | | Explosion | | | Damage (severity) | | | Estimated cost of repairs (£) | Estimated peak pressure (kN/m$^2$) |
|---|---|---|---|---|---|---|---|---|---|---|---|---|---|---|
| | | Use | Type (house only) | No of storeys | Age | Structure | Explosive substance | Source (piped gas only) | Ignition | Extent of significant damage | Fire† | Rating‡ (explosion only) | | |
| Jan–Mar 1972 (cont) | | | | | | | | | | | | | | |
| 1021 | 0 | E | – | 1 | a | B | S | – | L | 1 pumping station | B | S | 20 000 | – |
| 1032 | 0 | D | T | 2 | U | U | T | A | U | 3 houses | Nil | S | 2 000 | – |
| 1034 | 0 | D | T | 2 | b | B | T | E | L | 1 house | U | M | 1 500 | – |
| 1066 | 0 | D | U | 2 | U | U | T | A | U | 1 house | Nil | M | 500 | – |
| April–June 1972 | | | | | | | | | | | | | | |
| 1132 | 0 | D | T | 2 | a | B | N | E | L | 1 house | Nil | M | 2 500 | 5–15 |
| 1204 | 0 | D | T | 2 | a | B | N | U | L | 3 houses | U | VS | 10 000 | – |
| 1213 | 0 | I | – | 5 | a | B | O | – | A | 1 laboratory | A | M | 1 200 | 5–10 |
| 1236 | 0 | S/(F) | – | 3 | b | B | T | A | L | 1 shop | Nil | M | 500 | 2.5–7.5 |
| 1237 | 0 | F | – | 2 | a | B | T | A | A | 1 flat | Nil | M | 1 000 | 2.5–7.5 |
| 1255 | 1 | D | T | 2 | a | B | N | E | L | 4 houses | Nil | VS | 20 000 | – |
| 1266 | 0 | F | – | 4 | a | B | N | U | L | 1 flat | Nil | M | 2 500 | 2.5–7.5 |
| 1279 | 0 | D | U | U | c | U | N | A | U | 2 houses | A | S | 7 500 | – |
| 1356 | 0 | D | B | 1 | c | B | N | A | E | 1 bungalow | Nil | M | 1 000 | – |
| July–Sept 1972 | | | | | | | | | | | | | | |
| 1445 | 1 | D | T | 2 | a | B | T | I | U | 6 houses | Nil | S | 2 000 | – |
| 1448 | 0 | D | T | 2 | c | B | N | I | U | 4 houses | Nil | VS | 25 000 | – |
| 1539 | 0 | D | U | 2 | c | B | T | U | U | 1 house | A | M | 750 | – |
| 1601 | 0 | D | T | 3 | a | M | N | E | U | 1 house | Nil | M | 1 500 | – |
| 1604 | 0 | D | S | 2 | a | B | T | I | U | 1 house | Nil | S | 4 000 | 10–15 |
| 1613 | 0 | D | S | 2 | c | B | T | I | L | 1 house | U | S | 5 000 | – |
| 1630 | 0 | F | – | 3 | a | B | T | A | U | 1 flat | Nil | M | 2 500 | 5–15 |
| 1636 | 0 | D | S | 2 | c | B | T | A | L | 1 house | U | S | 1 000 | – |
| Oct–Dec 1972 | | | | | | | | | | | | | | |
| 1660 | 0 | S | – | 3 | c | U | T | E | E | 2 shops | Nil | S | 5 000 | – |
| 1664 | 0 | D | S | 2 | b | B | N | E | L | 1 house | A | M | 6 000 | – |
| 1669 | 0 | D | U | 2 | U | U | N | A | U | 1 house | U | M | 500 | – |
| 1689 | 0 | S/(F) | – | 2 | a | B | T | E | U | 2 shops/2 flats | Nil | S | 15 000 | – |
| 1749 | 0 | D | S | 2 | c | B | C | – | E | 1 house | A | M | 500 | – |
| 1755B | 0 | S | – | 2 | a | M | N | U | L | 1 shop | Nil | M | 1 500 | – |
| 1827 | 1 | D | B | 1 | c | B | N | A | U | 1 bungalow | A | S | 3 000 | – |
| 1855 | 0 | D | T | 2 | c | B | N | I | E | 1 house | Nil | M | 4 000 | – |
| 1864 | 0 | I | – | 1 | U | T | N | A | L | Outbuilding and classroom | Nil | M | 500 | – |
| 1873 | 0 | F | – | 2 | b | B | W | – | U | 1 flat | Nil | M | 500 | – |

| Incident no | No of fatal casualties | Building | | | | | Explosion | | | Damage (severity) | | | Estimated cost of repairs (£) | Estimated peak pressure (kN/m$^2$) |
|---|---|---|---|---|---|---|---|---|---|---|---|---|---|---|
| | | Use | Type (house only) | No of storeys | Age | Structure | Explosive substance | Source (piped gas only) | Ignition | Extent of significant damage | Fire† | Rating‡ (explosion only) | | |
| Oct–Dec 1972 (cont) | | | | | | | | | | | | | | |
| 1877 | 0 | S/(D) | T | 3 | a | B | P | – | F | 1 shop/2 houses | B | S | 1 500 | – |
| 1881 | 0 | D | T | 2 | U | U | N | E | U | 1 house | Nil | M | 1 000 | – |
| 1884 | 0 | D | D | 2 | a | M | U | U | U | 1 house | U | M | 750 | – |
| 1912 | 0 | F | – | U | U | U | N | U | U | 1 flat | U | M | 1 500 | – |
| Jan–Mar 1973 | | | | | | | | | | | | | | |
| 1936 | 0 | D | T | U | a | B | N | E | L | 7 houses | Nil | VS | 25 000 | – |
| 1978 | 0 | F | – | 3 | b | B | T | U | A | 1 flat | Nil | M | 750 | – |
| 2003 | 0 | D | T | 2 | a | B | N | I | A | 1 house | Nil | M | 500 | – |
| 2004 | 1 | D | T | 2 | b | B | T | I | U | 1 house | Nil | S | 1 000 | – |
| 2013 | 0 | D | B | 1 | c | B | N | I | U | 1 bungalow | Nil | S | 6 000 | – |
| 2036 | 0 | D | S | 2 | U | U | T | A | U | 1 house | Nil | S | 2 000 | – |
| 2092 | 0 | D | T | 2 | a | B | N | E | U | 1 house | A | S | 3 000 | – |
| 2099 | 0 | F/(S) | – | 3 | a | M | N | E | U | 1 shop/1 flat | Nil | M | 1 500 | 5–10 |
| 2201 | 0 | F | – | 3 | b | B | T | U | A | 1 flat | Nil | S | 7 500 | 10–15 |
| 2210 | 0 | D | T | 2 | a | B | T | U | U | 2 houses | Nil | S | 1 500 | – |
| April–June 1973 | | | | | | | | | | | | | | |
| 14 | 0 | D | U | U | U | U | C | – | F | 1 house | B | M | 1 000 | – |
| 63 | 0 | D | 1S/2T | 3 | a | B | N | E | U | 3 houses | A | VS | 30 000 | – |
| 89 | 0 | I | – | 1 | c | B | T | E | U | 1 classroom | A | M | 750 | – |
| 100 | 0 | E | – | 1 | U | B/C | S | – | U | 1 pumping station | Nil | S | 5 000 | – |
| 137 | 0 | D | U | U | U | U | N | U | L | 1 house | A | M | 750 | – |
| 142 | 0 | D | S | 2 | b | B | C | – | F | 1 house | B | M | 2 500 | – |
| 178 | 0 | D | B | 1 | U | U | C | – | U | 1 bungalow | Nil | M | 6 000 | – |
| 192 | 0 | D | T | 2 | a | M | W | – | – | 1 house | Nil | S | 3 000 | – |
| 207 | 0 | D | S | 2 | c | B | N | I | U | 2 houses | Nil | S | 7 500 | – |
| 233 | 0 | S | – | U | U | U | N | E | U | 1 public house | Nil | M | 1 250 | – |
| July–Sept 1973 | | | | | | | | | | | | | | |
| 248 | 0 | D | D | 2 | a | B | C | – | E | 1 house | B | M | 3 000 | – |
| 271 | 0 | D | U | U | U | U | W | – | – | 1 house | Nil | M | 750 | – |
| 312 | 0 | D | T | U | U | U | N | E | U | 2 houses | Nil | M | 2 000 | – |
| 344 | 0 | D | S | 2 | c | T | N | E | L | 2 houses | A | S | 12 000 | – |
| 348 | 1 | D | D | 2 | c | B | T | I | U | 1 house | A | S | 12 000 | – |
| 372 | 0 | X | – | U | U | U | N | E | A | 1 clubhouse | A | M | 2 000 | – |
| 375 | 0 | W | – | 1 | U | B/S | P | – | U | 1 workshop | B | S | 4 500 | – |
| 387 | 0 | W | – | 1 | U | C | C | – | F | 1 store | B | S | 5 000 | – |

| Incident no | No of fatal casualties | Building | | | | | Explosion | | | Damage (severity) | | | Estimated cost of repairs (£) | Estimated peak pressure (kN/m²) |
|---|---|---|---|---|---|---|---|---|---|---|---|---|---|---|
| | | Use | Type (house only) | No of storeys | Age | Structure | Explosive substance | Source (piped gas only) | Ignition | Extent of significant damage | Fire† | Rating‡ (explosion only) | | |
| Oct–Dec 1973 | | | | | | | | | | | | | | |
| 412 | 0 | D | T | 2 | a | B | W | – | – | 2 houses | Nil | M | 2 500 | – |
| 437 | 0 | D | 2S/2T | 2 | b | B | N | E | U | 4 houses | A | S | 21 000 | – |
| 461 | 0 | D | S | 2 | a | M | N | E | L | 1 house | A | M | 750 | – |
| 463 | 1 | D | T | 2 | a | B | N | E | U | 2 houses | A | S | 7 500 | – |
| 465 | 0 | E | – | 1 | U | B/C | U | U | E | 1 sub-station | Nil | S | 7 000 | – |
| 492 | 2 | S/(F) | – | 3 | a | B | N | E | U | 3 flats/3 shops | A | VS | 30 000 | – |
| 498 | 0 | D | D | 2 | b | B | T | E | U | 1 house | Nil | M | 2 500 | – |
| 509 | 0 | D | S | 2 | b | B | N | U | U | 1 house | A | M | 1 250 | – |
| 522 | 0 | D | D | 2 | a | U | T | U | U | 1 house | A | M | 1 750 | – |
| 531 | 0 | F | – | U | U | U | C | – | U | 1 flat | B | M | 500 | – |
| 532 | 0 | D | S | 2 | a | B | N | A | L | 1 house | Nil | M | 1 000 | – |
| 565 | 0 | D | B | 1 | U | U | N | A | U | 1 bungalow | A | M | 1 250 | – |
| 572 | 0 | D | S | 2 | b | B | T | A | U | 1 house | Nil | S | 7 500 | – |
| 578 | 0 | D | U | U | U | U | N | E | U | 1 house | Nil | M | 1 000 | – |
| 582 | 0 | F | – | U | U | U | T | U | U | 1 flat | A | M | 1 000 | – |
| Jan–Mar 1974 | | | | | | | | | | | | | | |
| 604 | 0 | F | – | 7 | c | B/CF | N | E | A | 2 flats | A | M | 750 | – |
| 605 | 0 | D | S | 2 | U | U | N | A | U | 1 house | A | M | 750 | – |
| 627 | 0 | W/(D) | S | 2 | a | B | N | E | E | 1 store/2 houses | Nil | S | 6 000 | – |
| 644 | 0 | D | S | 2 | c | C | C | – | E | 1 house | Nil | M | 2 500 | – |
| 670 | 0 | D | S | 2 | b | B | N | E | L | 1 house | A | M | 1 000 | – |
| 700 | 1 | D | B | 1 | c | B | N | A | L | 1 bungalow | A | M | 4 000 | – |
| 706 | 0 | D | T | 2 | U | U | N | I | U | 1 house | Nil | M | 750 | – |
| 710 | 0 | D | S | 2 | a | B | N | I | U | 1 house | A | S | 2 500 | – |
| 720 | 0 | D | S | 2 | b | B | N | E | U | 1 house | A | M | 2 500 | – |
| 723 | 0 | D | T | 2 | a | B | N | E | L | 2 houses | Nil | S | 7 000 | – |
| 745 | 0 | D | S | 2 | b | B | N | I | F | 2 houses | B | S | 5 000 | – |
| 752 | 1 | F | – | 2 | C | B | N | A | U | 1 flat | A | M | 1 000 | – |
| April–June 1974 | | | | | | | | | | | | | | |
| 800 | 0 | F | – | 10 | c | B/CF | N | A | U | 1 flat | Nil | M | 1 500 | – |
| 813 | 0 | D | D | 2 | a | M | U | U | U | 1 house | A | S | 7 000 | – |
| 846 | 3 | D | D | 2 | a | M | N | E | U | 1 house | A | S | 8 500 | – |
| 852 | 1 | D | T | 2 | a | B | T | I | L | 3 houses | Nil | S | 20 000 | – |
| 861 | 0 | D | T | 2 | a | M | T | E | U | 1 house | A | M | 4 000 | – |
| 863 | 0 | W | – | 1 | C | B | C | – | F | 3 garages | B | M | 1 250 | – |

| Incident no | No of fatal casualties | Building | | | | | Explosion | | | Damage (severity) | | | Estimated cost of repairs (£) | Estimated peak pressure (kN/m$^2$) |
|---|---|---|---|---|---|---|---|---|---|---|---|---|---|---|
| | | Use | Type (house only) | No of storeys | Age | Structure | Explosive substance | Source (piped gas only) | Ignition | Extent of significant damage | Fire† | Rating‡ (explosion only) | | |
| April–June 1974 (cont) | | | | | | | | | | | | | | |
| 873 | 0 | D | D | 2 | U | U | X | – | – | 1 house | Nil | M | 1 500 | – |
| 881 | 0 | D | S | 2 | c | B | N | I | E | 1 house | A | M | 1 250 | – |
| 883 | 0 | D | T | 2 | c | CW | N | A | L | 1 house | Nil | M | 1 500 | – |
| 897 | 0 | D | U | U | U | U | N | U | U | 1 house | Nil | M | 2 500 | – |
| 918 | 1 | D | S | 2 | b | B | N | I | U | 2 houses | Nil | S | 17 000 | – |
| 976 | 0 | D | S | 2 | U | B | U | U | U | 1 house | Nil | M | 2 000 | – |
| July–Sept 1974 | | | | | | | | | | | | | | |
| 995 | 0 | D | U | U | U | U | T | A | L | 1 house | Nil | M | 1 500 | – |
| 1020 | 1 | D | U | U | U | B | C | – | F | 1 house | B | M | 750 | – |
| 1025 | 0 | H | – | 3 | a | M | C | – | F | 1 hotel | B | M | 1 000 | – |
| 1040 | 0 | S | – | 2 | U | B/C | N | E | U | 1 shop | Nil | M | 500 | – |
| 1065 | 1 | F | – | 3 | b | B | T | A | U | 3 flats | Nil | S | 12 500 | – |
| 1074 | 0 | S/(F) | – | 2 | a | B | N | E | A | 1 flat/1 shop | Nil | S | 10 000 | – |
| 1089 | 0 | F | – | 2 | c | B | N | I | L | 4 flats | Nil | S | 25 000 | – |
| 1115 | 0 | D | D | 3 | b | B | N | E | U | 1 house | Nil | M | 3 000 | – |
| 1117 | 0 | D | S | 2 | c | B | T | U | U | 1 house | Nil | M | 2 000 | – |
| 1133 | 0 | D | S | 2 | b | B | T | I | U | 1 house | Nil | M | 750 | – |
| 1143 | 0 | D | T | U | U | U | T | I | L | 1 house | A | M | 1 000 | – |
| 1149 | 0 | I | – | 1 | c | B | C | – | F | 1 boilerhouse | Nil | M | 1 000 | – |
| 1151 | 1 | D | S | 2 | b | B | N | E | L | 2 houses | Nil | S | 20 000 | – |
| 1189 | 0 | D | T | 2 | a | B | N | I | E | 1 house | Nil | S | 6 000 | – |
| Oct–Dec 1974 | | | | | | | | | | | | | | |
| 1249 | 0 | W | – | 1 | c | B/S | C | – | L | 1 garage | A | M | 1 250 | – |
| 1251 | 0 | D | U | 2 | U | U | N | A | L | 1 house | Nil | M | 1 000 | – |
| 1280 | 0 | T/(W) | – | 1 | b | B/C/S | N | E | E | 1 tel exchange/1 garage | Nil | VS | 40 000 | 20–35 |
| 1285 | 0 | D | S | 2 | c | B | N | I | U | 2 houses | A | VS | 20 000 | – |
| 1295 | 0 | D | S | 2 | c | B | N | E | U | 1 house | A | M | 500 | – |
| 1298 | 0 | D | B | 1 | c | B | N | I | L | 1 bungalow | Nil | M | 1 750 | – |
| 1322 | 0 | X | – | 1 | a | M | C | – | E | 1 chapel | Nil | M | 2 500 | – |
| 1361 | 0 | D | B | 1 | c | B | N | E | U | 1 bungalow | Nil | M | 15 000 | – |
| 1368 | 0 | D | S | 3 | a | B | N | E | U | 1 house | Nil | M | 1 750 | – |
| 1379 | 0 | D | T | U | U | B | N | I | E | 1 house | Nil | M | 750 | – |
| 1396 | 0 | D | S | 2 | a | M | N | E | F | 2 houses | B | S | 17 500 | – |
| 1444 | 0 | D | U | U | U | U | C | – | U | 1 house | Nil | M | 2 000 | – |
| 1454 | 0 | S/(F) | – | 2 | U | B | W | – | – | 1 shop/1 flat | Nil | S | 6 000 | – |

| Incident no | No of fatal casualties | Building | | | | | Explosion | | | Damage (severity) | | | Estimated cost of repairs (£) | Estimated peak pressure (kN/m$^2$) |
|---|---|---|---|---|---|---|---|---|---|---|---|---|---|---|
| | | Use | Type (house only) | No of storeys | Age | Structure | Explosive substance | Source (piped gas only) | Ignition | Extent of significant damage | Fire† | Rating‡ (explosion only) | | |
| Jan–Mar 1975 | | | | | | | | | | | | | | |
| 1501 | 0 | S/(F) | – | 2 | a | B | P | – | L | 3 flats/3 shops | Nil | S | 22 500 | – |
| 1505 | 0 | D | S/3T | 2 | C | B | N | E | U | 4 houses | A | VS | 25 000 | – |
| 1523 | 0 | D | U | U | C | B | N | E | U | 1 house | Nil | M | 1 250 | – |
| 1553 | 0 | F | – | 3 | a | B | W | – | – | 2 flats | Nil | M | 2 000 | – |
| 1556 | 0 | S/(F) | – | 7 | c | B/CF | N | E | A | 2 flats/2 shops | Nil | M | 4 000 | – |
| 1574 | 0 | F | – | 2 | a | B | N | A | L | 1 flat | A | M | 2 500 | – |
| 1587 | 0 | D | S | 2 | U | B | N | I | U | 1 house | Nil | M | 1 000 | – |
| 1595 | 0 | O | – | U | U | U | N | I | U | 1 office | Nil | M | 3 000 | – |
| 1598 | 1 | D | T | 2 | a | B | N | E | L | 2 houses | A | S | 12 500 | – |
| 1643 | 0 | D | T | 2 | U | B | N | E | U | 1 house | Nil | S | 9 000 | – |
| 1649 | 0 | O/(W) | – | U | U | U | N | E | A | 1 office/1 store | A | M | 2 500 | – |
| 1667 | 0 | D | T | 2 | U | B | T | I | L | 1 house | Nil | M | 4 000 | – |
| April–June 1975 | | | | | | | | | | | | | | |
| 1713 | 0 | D | U | 2 | U | B | C | – | U | 1 house | Nil | M | 750 | – |
| 1757 | 0 | F | – | 5 | c | U | N | A | L | 1 flat | A | M | 3 500 | – |
| 1778 | 0 | D | T | 2 | a | B | N | I | U | 1 house | Nil | M | 5 000 | – |
| 1784 | 0 | D | S | 2 | a | B | N | U | L | 1 house | Nil | M | 4 000 | – |
| 1826 | 0 | D | U | U | U | U | N | U | U | 1 house | Nil | M | 750 | – |
| 1827 | 0 | S | – | U | U | U | U | U | F | 1 shop | B | M | 2 000 | – |
| 1829 | 0 | D | S/2T | 2 | c | B | N | I | U | 3 houses | Nil | S | 30 000 | – |
| 1847 | 0 | W | – | 1 | U | M/S | C | – | F | 1 store | B | M | 4 000 | – |
| July–Sept 1975 | | | | | | | | | | | | | | |
| 1912 | 0 | W | – | 1 | U | B | C | – | L | 1 store | A | M | 3 500 | – |
| 1996 | 0 | F | – | 15 | c | CW/CF | N | I | U | 4 flats | Nil | VS | 100 000 | 30–40 (50–100)* |
| 2019 | 0 | D | T | 2 | c | S | N | I | U | 1 house | Nil | M | 4 000 | – |
| 2051 | 0 | F | – | 4 | U | U | N | U | U | 1 flat | A | M | 1 500 | – |
| 2081 | 0 | X | – | 2 | c | T | W | – | – | 1 club | Nil | M | 7 500 | – |
| Oct–Dec 1975 | | | | | | | | | | | | | | |
| – | 0 | F | – | 4 | a | B | N | A | L | 1 flat | A | M | 7 500 | – |
| 2119 | 1 | D | S | 2 | a | B | N | A | L | 1 house | Nil | M | 2 500 | – |
| 2129 | 0 | D | D | 2 | U | U | N | E | U | 1 house | Nil | S | 6 000 | – |
| 2145 | 0 | D | T | 2 | U | U | N | I | U | 1 house | A | M | 2 250 | – |

*Secondary peak

| Incident no | No of fatal casualties | Building | | | | | Explosion | | | Damage (severity) | | | Estimated cost of repairs (£) | Estimated peak pressure (kN/m²) |
|---|---|---|---|---|---|---|---|---|---|---|---|---|---|---|
| | | Use | Type (house only) | No of storeys | Age | Structure | Explosive substance | Source (piped gas only) | Ignition | Extent of significant damage | Fire† | Rating‡ (explosion only) | | |
| Oct–Dec 1975 (cont) | | | | | | | | | | | | | | |
| 2152 | 0 | D | S | 2 | b | B | N | I | A | 2 houses | Nil | M | 3 000 | – |
| 2170 | 0 | D | T | 2 | a | B | N | I | E | 3 houses | Nil | S | 25 000 | – |
| 2219 | 0 | D | D | 2 | a | B | N | E | U | 1 house | A | M | 2 000 | – |
| 2226 | 1 | D | S/T | 2 | a | B | N | E | U | 2 houses | Nil | VS | 25 000 | – |
| 2261 | 0 | O/(S) | – | 11 | c | B/CF | C | – | F | 1 office/2 shops | B | S | 6 000 | – |
| 2265 | 0 | I | – | 1 | c | B | N | U | E | 1 laboratory | Nil | M | 6 000 | – |
| 2303 | 0 | D | S | 2 | c | B | N | E | L | 1 house | Nil | M | 750 | – |
| 2308 | 0 | D | D | 2 | U | U | N | E | U | 1 house | A | M | 7 000 | – |
| 2309 | 0 | F | – | 3 | U | B | T | I | U | 1 flat | A | M | 2 500 | – |
| 2314 | 0 | F | – | 3 | c | B | N | I | U | 1 flat | A | M | 750 | – |
| Jan–Mar 1976 | | | | | | | | | | | | | | |
| 2343 | 0 | D | 1S/1T | 2 | c | T | N | I | U | 2 houses | Nil | S | 12 000 | – |
| 2381 | 0 | D | U | U | U | U | N | A | L | 1 house | A | M | 1 500 | – |
| 2410 | 0 | D | T | U | a | B | N | I | U | 1 house | Nil | M | 1 500 | – |
| 2414 | 0 | D | B | 1 | U | B | N | I | U | 1 bungalow | A | M | 1 000 | – |
| 2438 | 0 | D | U | 2 | U | U | N | E | L | 1 house | Nil | M | 1 250 | – |
| 2441 | 0 | D | S | 2 | U | B | N | E | U | 1 house | Nil | M | 2 500 | – |
| 2449 | 0 | S/(F) | – | 2 | a | B | W | – | – | 1 flat/1 shop | Nil | M | 3 000 | – |
| 2479 | 0 | F | – | 5 | b | B | N | E | E | 2 flats | Nil | M | 5 000 | – |
| 2480 | 0 | D | T | 2 | a | B | N | I | U | 1 house | Nil | M | 1 500 | – |
| 2488 | 0 | D | T | 2 | a | B | N | E | U | 1 house | Nil | M | 5 000 | – |
| 2497 | 0 | D | S | 2 | a | B | N | E | U | 1 house | A | M | 10 000 | – |
| 2576 | 0 | D | S | 2 | b | B | N | I | U | 1 house | Nil | S | 8 000 | – |
| 2603 | 1 | D | T | 2 | b | B | N | A | U | 1 house | Nil | M | 1 500 | – |
| 2606 | 0 | O | – | 1 | U | C | C | – | U | 1 office | Nil | S | 3 000 | – |
| 2610 | 0 | D | B | 1 | U | U | N | E | U | 1 bungalow | A | M | 1 000 | – |
| 2621 | 1 | D | S | 2 | c | S | N | I | L | 1 house | A | M | 6 000 | – |
| 2642 | 0 | D | T | 2 | a | B | N | I | E | 2 houses | B | S | 8 000 | – |
| 2646 | 0 | D | B | 1 | c | B | N | A | U | 1 bungalow | Nil | S | 7 500 | – |
| 2648 | 0 | D | S | 2 | b | B | C | – | U | 1 house | Nil | M | 5 000 | – |
| April–June 1976 | | | | | | | | | | | | | | |
| 2662 | 0 | D | T | 2 | a | B | N | I | U | 1 house | Nil | S | 7 500 | – |
| 2679 | 0 | D | T | 2 | U | U | N | E | U | 1 house | Nil | M | 1 000 | – |
| 2683 | 0 | F | S | 2 | c | B | N | I | U | 2 flats | Nil | S | 5 000 | – |
| 2712 | 0 | D | U | U | U | U | W | – | – | 1 house | Nil | M | 750 | – |
| 2743 | 0 | F | T | 2 | c | B | N | I | F | 4 flats | Nil | S | 40 000 | 15–30 |

| Incident no | No of fatal casualties | Building | | | | | Explosion | | | Damage (severity) | | | Estimated cost of repairs (£) | Estimated peak pressure (kN/m$^2$) |
|---|---|---|---|---|---|---|---|---|---|---|---|---|---|---|
| | | Use | Type (house only) | No of storeys | Age | Structure | Explosive substance | Source (piped gas only) | Ignition | Extent of significant damage | Fire† | Rating‡ (explosion only) | | |
| April–June 1976 (cont) | | | | | | | | | | | | | | |
| 2773 | 0 | D | B | 1 | c | B | C | – | U | 1 bungalow | Nil | M | 10 000 | – |
| 2774 | 0 | D | S | 2 | U | B | X | – | – | 1 house | Nil | M | 750 | – |
| 2779 | 1 | D | T | 2 | a | B | N | I | L | 2 houses | Nil | S | 12 500 | – |
| 2786 | 0 | D | S | 2 | c | CW | C | – | – | 1 house | A | M | 750 | – |
| 2805 | 0 | D | B | 1 | c | B | N | A | U | 1 bungalow | A | M | 7 500 | – |
| 2847 | 0 | E | – | 1 | U | B | U | U | E | 1 sub-station | Nil | M | 750 | – |
| July–Sept 1976 | | | | | | | | | | | | | | |
| 2954 | 1 | D | T | U | U | B | N | I | L | 1 house | Nil | M | 2 000 | – |
| 2989 | 0 | D | T | 2 | a | B | N | I | L | 1 house | Nil | M | 1 000 | – |
| 3005 | 0 | S | – | 1 | c | B | C | – | – | 1 restaurant | Nil | M | 2 000 | – |
| 3015 | 0 | E | – | 1 | c | B | U | U | E | 1 sub-station | Nil | M | 2 000 | – |
| 3086 | 0 | D | S | 2 | c | B | N | E | L | 2 houses | A | S | 30 000 | – |
| 3092 | 0 | F | – | 12 | c | B/CF | N | E | L | 1 flat | A | S | 2 500 | 25–35 |
| 3100 | 0 | D | b | 1 | c | B | N | E | U | 1 bungalow | Nil | S | 12 000 | – |
| 3101 | 0 | D | S | 2 | b | B | W | – | – | 1 house | Nil | S | 6 000 | (25–50)* |
| 3127 | 0 | D | S | 2 | a | B | N | U | L | 1 house | A | M | 3 000 | – |
| 3133 | 0 | D | B | 1 | b | B | N | I | L | 1 bungalow | Nil | M | 5 000 | – |
| Oct–Dec 1976 | | | | | | | | | | | | | | |
| 3187 | 0 | D | S | 2 | b | B | N | E | L | 1 house | Nil | S | 12 500 | – |
| 3193 | 1 | F | – | 3 | a | M | N | E | L | 9 flats | A | S | 60 000 | – |
| 3213 | 0 | S | – | 3 | a | B | C | – | U | 1 public house | Nil | M | 3 000 | – |
| 3240 | 1 | D | T | 2 | a | B | N | A | U | 1 house | Nil | M | 1 250 | – |
| 3248 | 0 | F | – | 3 | c | B/M | N | I | E | 4 flats | Nil | S | 35 000 | 15–30 |
| 3252 | 0 | D | S | 2 | a | B | N | E | L | 1 house | Nil | S | 15 000 | – |
| 3296 | 0 | I | – | U | c | U | N | I | F | 1 boilerhouse | B | M | 2 000 | – |
| 3303 | 0 | D | D | U | U | U | L | U | U | 1 house | Nil | S | 25 000 | – |
| 3310 | 0 | D | T | 3 | a | B | N | E | E | 1 house | Nil | M | 1 250 | – |
| 3344 | 1 | D | S | 2 | a | B | N | E | L | 1 house | Nil | S | 10 000 | – |
| 3346 | 0 | E | – | 1 | U | B | U | U | E | 1 sub-station | Nil | M | 1 500 | – |
| 3348 | 0 | D | T | 3 | a | B | N | I | L | 1 house | A | M | 16 000 | – |
| 3356 | 0 | F | – | U | c | U | N | E | L | 1 flat | Nil | M | 1 500 | – |
| 3358 | 0 | F | – | 1 | c | B | C | – | U | 1 flat | B | M | 10 000 | – |
| 3372 | 0 | D | S | 2 | b | B | N | I | U | 1 house | Nil | S | 12 500 | – |
| 3373 | 0 | O/(S) | – | 7/5/5 | a | B/M | N | E | U | 3 office buildings/3 shops | A | VS | 300 000 | 35–70 |
| 3374 | 0 | S/(F) | – | 3 | b | B/M | N | E | E | 1 shop/2 flats | A | S | 15 000 | 20–35 |
| 3379 | 0 | X | – | 1 | U | U | C | – | U | 1 public convenience | Nil | M | 3 000 | – |

*Water heater

| Incident no | No of fatal casualties | Building | | | | | Explosion | | | Damage (severity) | | | Estimated cost of repairs (£) | Estimated peak pressure (kN/m²) |
|---|---|---|---|---|---|---|---|---|---|---|---|---|---|---|
| | | Use | Type (house only) | No of storeys | Age | Structure | Explosive substance | Source (piped gas only) | Ignition | Extent of significant damage | Fire† | Rating‡ (explosion only) | | |
| Jan–Mar 1977 | | | | | | | | | | | | | | |
| 3392 | 0 | F/(S) | – | 3 | a | B | N | I | A | 3 shops/3 flats | Nil | S | 60 000 | 20–35 |
| 3393 | 0 | W | – | 1 | c | S | O | – | F | 2 warehouses | B | S | 60 000 | – |
| 3394 | 1 | D | S | 2 | c | B/S | C | – | U | 2 houses | A | S | 20 000 | – |
| 3395 | 0 | W | – | 1 | c | B/S | C | – | U | 1 warehouse | Nil | S | 30 000 | – |
| 3404 | 0 | I | – | 1 | a | M | T | I | L | 1 hospital bakery | Nil | M | 4 000 | – |
| 3432 | 0 | D | T | 2 | a | B | N | E | U | 1 house | A | M | 2 000 | – |
| 3433 | 0 | D | B | 1 | c | B | N | I | L | 1 bungalow | A | M | 10 000 | – |
| 3435 | 0 | F | – | 3 | c | B | N | I | L | 2 flats | Nil | S | 20 000 | 10–20 |
| 3436 | 0 | D | T | 2 | a | B | N | E | U | 1 house | Nil | S | 7 500 | – |
| 3492 | 0 | D | T | 2 | a | B | N | E | L | 1 house | A | M | 1 500 | – |
| 3493 | 0 | D | B | 1 | c | B | N | I | U | 1 house | Nil | M | 7 500 | – |
| 3627 | 0 | D | S | 2 | c | B | N | I | F | 1 house | Nil | M | 2 000 | – |
| 3645 | 0 | D | S | 2 | b | B | N | I | L | 1 house | Nil | M | 1 500 | – |
| 3665 | 0 | D | D | 2 | a | M | N | E | L | 1 house | Nil | M | 1 500 | – |
| 3681 | 0 | S/(F) | – | 3 | b | B | N | I | F | 1 shop | B | M | 2 000 | – |
| 3684 | 0 | S | – | 1 | U | M | U | U | U | 1 public house | Nil | M | 5 000 | – |
| 3707 | 0 | D | T | 2 | a | B | N | I | L | 1 house | A | M | 1 250 | – |
| 3714 | 0 | F | – | 2 | a | B | N | E | E | 2 flats | Nil | M | 5 000 | – |
| 3719 | 0 | E | – | 1 | c | B/C | N | I | U | 1 sub-station | A | M | 2 500 | – |
| 3723 | 0 | S/(F) | – | 4 | b | B | N | I | F | 1 shop | B | M | 4 000 | – |

†B before; A after.

‡M moderate; S severe; VS very severe.

## Appendix B Details of all significant vehicle impacts

| Quarter | Incident no | No of fatal casualties* | Building(s) affected | | | | | Vehicle type | Extent of significant damage | Rating† | Estimated cost of repairs (£) |
|---|---|---|---|---|---|---|---|---|---|---|---|
| | | | Use | Type (house only) | No of storeys | Age | Structure | | | | |
| April–June 1971 | 92 | 0 | D | T | 2 | a | B | Lorry | 2 walls | M | 400 |
| | 97 | 0 | D | U | U | a | U | Tanker | 1 wall | M | 350 |
| | 114 | 0 | D | B | 1 | c | M | Lorry | 1 corner | M | 500 |
| | 120 | 0 | D | T | U | U | U | Lorry | Front wall | M | 250 |
| | 131 | 0 | D | B | 1 | c | B | Van | Front wall | M | 750 |
| | 141 | 0 | D | S/T | 2 | a | U | Car | 1 wall | M | 350 |
| | 142 | 0 | D | U | 2 | b/c | U | Lorry | Front wall | M | 750 |
| | 143 | 0 | D | U | U | a | U | Lorry | Kitchen wall | M | 400 |
| | 144 | 0 | D | S | 2 | c | B | Lorry | Front wall | M | 500 |
| | 147D | 0 | D | T | 2 | b | B | Lorry | 1 wall | M | 350 |
| July–Sept 1971 | 166 | 0 | S | – | U | U | U | Coach | 1 wall | M | 400 |
| | 179 | 0 | D | T | 2 | a | B | Lorry | Front wall | M | 400 |
| | 185 | 0 | D | B | 1 | b | T | Fire engine | 1 ext/1 int wall | M | 500 |
| | 195 | 0 | D | T | 2 | a | B | Van | Front wall | M | 250 |
| | 200 | 0 | D | S | 2 | b | B | Bus | Kitchen wall | M | 350 |
| | 201 | 0 | D | U | U | U | U | Coach | Front wall | M | 250 |
| | 238 | 0 | D | U | U | b | U | Ambulance | Kitchen wall | M | 200 |
| | 241 | 0 | S | – | 2 | a | B | Lorry | Front wall | M | 400 |
| | 265 | 8(V) | D | T | 2 | a | M | Car & coach | Front wall/int walls | M | 750 |
| | 268 | 0 | D | T | 3/4 | a | B | Lorry | Front wall | M | 250 |
| | 283 | 0 | D | U | U | U | U | Lorry | Front wall | M | 200 |
| | 302 | 0 | D | T | 2 | a | M | Lorry | Side wall | M | 300 |
| | 309 | 0 | D | U | U | b/c | U | Lorry | Front wall | M | 200 |
| | 319 | 0 | D | S | 2 | a | B | Coach | Front wall/floor/gable end | M | 750 |
| | 323 | 0 | D | S | 2 | a | B | Coach | Front wall | M | 600 |
| | 334 | 0 | D | D | 2 | a | B | Lorry | 1 wall | M | 600 |
| | 352 | 0 | D | U | U | b/c | B | Lorry | Front wall | M | 250 |
| | 368 | 0 | D | B | 1 | a | B | Car | Side wall | M | 300 |
| | 369 | 0 | D | U | U | b | B | Car | Front wall | M | 200 |
| | 372 | 0 | D | T | 2 | a | B | Lorry | Entire front wall | M | 1 000 |
| | 377 | 0 | D | T | 2 | b | B | Coach | Front wall | M | 450 |
| Oct–Dec 1971 | 386 | 0 | D | T | 2 | a | B | Lorry | Front wall | M | 500 |
| | 388 | 0 | D | U | U | c | B | Lorry | Side wall/int walls | M | 1 000 |
| | 400 | 0 | D | U | 2 | a | B | Lorry | Front wall | M | 500 |

*(V) in vehicle; (B) in building; (S) in street.
†M moderate; S severe.

| Quarter | Incident no | No of fatal casualties* | Building(s) affected | | | | | Vehicle type | Extent of significant damage | Rating† | Estimated cost of repairs (£) |
|---|---|---|---|---|---|---|---|---|---|---|---|
| | | | Use | Type (house only) | No of storeys | Age | Structure | | | | |
| Oct– Dec 1971 (cont) | 414 | 0 | D | U | 2 | U | U | Lorry | Front wall | M | 250 |
| | 416 | 1(V) | S | – | U | a | B | Van | Side wall | M | 350 |
| | 436 | 0 | D | B | 1 | c | B | Truck | Side wall | M | 300 |
| | 453 | 0 | D | U | 2 | U | U | Car & van | Front wall | M | 500 |
| | 460 | 0 | D | U | 2 | a | B | Tanker | Front wall | M | 400 |
| | 493 | 0 | D | S | 2 | a | B | Lorry | Front wall | M | 250 |
| | 506I | 0 | D | U | U | a | B | Car | 1 wall/outhouse | M | 500 |
| | 554 | 0 | D | U | 2 | c | U | Lorry | 1 wall | M | 250 |
| | 580 | 0 | D | B | 1 | c | B | Car | Front wall | M | 250 |
| | 594 | 0 | 2D | T | 2 | a | B | Fire engine | 2 front walls/party wall | M | 750 |
| | 595 | 0 | D | U | U | U | U | Tanker | 1 wall | M | 750 |
| | 613 | 0 | D | U | U | U | U | Lorry | Front wall | M | 350 |
| | 634 | 0 | D | T | 2 | a | B | Car | Front wall | M | 350 |
| | 665 | 0 | D | T | 2 | c | B | Bulldozer | Front wall | M | 250 |
| | 681 | 0 | D | B | 1 | b | T | Car | 2 walls | M | 750 |
| | 691 | 0 | D | U | U | a | B | Lorry | Front wall | M | 250 |
| | 712 | 0 | D | U | U | a | U | Car | Front wall | M | 350 |
| Jan– Mar 1972 | 719 | 0 | D | U | U | c | B | Car | Garage wall | M | 400 |
| | 726 | 0 | D | S | 2 | b | B | Lorry | Front wall | M | 600 |
| | 731 | 0 | S | – | 1 | a | U | Lorry | 1 ext/1 int wall | M | 750 |
| | 756 | 0 | D | U | U | U | U | Lorry | 1 wall | M | 750 |
| | 811 | 0 | D | S | 2 | a | B | Lorry | Side wall | M | 300 |
| | 870 | 0 | D | U | U | a | B | Lorry | Front wall | M | 250 |
| | 887 | 0 | D | U | U | a | U | Lorry | 1 wall | M | 350 |
| | 888 | 0 | D | U | 2 | a | B | Bus | Front wall | M | 350 |
| | 915 | 0 | D | S | 2 | a | B | Bus | 2 ext walls | M | 750 |
| | 973 | 0 | D | U | U | a | U | Lorry | Front wall | M | 450 |
| | 1030 | 0 | D | U | U | a | B | Car | Front wall | M | 300 |
| | 1067 | 1(V) | I | – | U | U | U | Lorry | Side wall | M | 500 |
| | 1073 | 1(V) | D | T | 2 | a | B | Car | Front wall | M | 300 |
| | 1078 | 1(B) | D | U | 2 | a | U | Lorry | Front wall/int wall | M | 1 000 |
| | 1089 | 0 | I | – | U | U | B | Van | 1 classroom | M | 250 |
| | 1108 | 0 | D | U | U | a | U | Lorry | Front wall | M | 250 |
| | 1118 | 0 | D | U | U | a | U | Lorry | Side wall | M | 350 |

*(V) in vehicle; (B) in building; (S) in street.
†M moderate; S severe.

| Quarter | Incident no | No of fatal casualties* | Building(s) affected | | | | | Vehicle type | Extent of significant damage | Rating† | Estimated cost of repairs (£) |
|---|---|---|---|---|---|---|---|---|---|---|---|
| | | | Use | Type (house only) | No of storeys | Age | Structure | | | | |
| April–June 1972 | 1158 | 0 | S | – | U | a | U | Lorry | Front wall | M | 300 |
| | 1163 | 1(V) | D | B | 1 | a | U | Car | Side wall | M | 500 |
| | 1177 | 0 | D | U | U | a | U | Steam roller | Front wall | M | 250 |
| | 1218 | 0 | D | U | 2 | a | B | Lorry | Side wall | M | 450 |
| | 1232 | 0 | D | U | 2 | a | U | Excavator | 1 wall | M | 350 |
| | 1239 | 0 | S | – | U | a | B | Car | Front wall | M | 250 |
| | 1281 | 0 | 2D | T | U | U | U | Lorry | 2 front walls | M | 600 |
| | 1316 | 0 | D | S | 2 | a | B | Bus | 2 walls/roof/gable end | M | 1 750 |
| | 1325 | 0 | D | S | 2 | a | B | Lorry | Front wall | M | 250 |
| | 1329 | 0 | W | – | 1 | a | B | Car | 1 wall | M | 350 |
| | 1385 | 0 | D | U | U | c | B | Lorry | Front wall | M | 250 |
| | 1392 | 0 | S | – | 3 | a | U | Lorry | Front wall | M | 600 |
| July–Sept 1972 | 1402 | 0 | D | U | 2 | c | B | Car | Side wall | M | 350 |
| | 1408 | 0 | D | S | 2 | a | B | Lorry | Front wall | M | 400 |
| | 1439 | 0 | W | – | 1 | U | U | Lorry | 1 wall | M | 600 |
| | 1450 | 0 | D | B | 1 | U | U | Lorry | 1 wall | M | 500 |
| | 1467 | 0 | D | U | U | a | U | Lorry | Front wall | M | 500 |
| | 1477 | 0 | S | – | 2 | a | B | Lorry | Front wall | M | 600 |
| | 1498 | 0 | 2D | T | 2 | a | B | Lorry | 2 front walls | M | 1 000 |
| | 1508 | 0 | S/F | – | 2 | b | B | Lorry | 2 walls/floor | S | 2 000 |
| | 1517 | 0 | D | U | U | a | U | Lorry | 1 wall | M | 500 |
| | 1527 | 0 | W | – | 1 | a | B | Lorry | 1 wall/roof | M | 500 |
| | 1546 | 0 | D | U | 2 | a | B | Bulldozer | 1 wall | M | 750 |
| | 1568 | 0 | S | – | 2 | a | B | Lorry | 1 ext/1 int wall | M | 750 |
| | 1612 | 0 | D | U | 2 | a | B | Van | Front wall | M | 400 |
| | 1620 | 0 | S/F | – | 2 | b | B | Lorry | Front wall/upper floor | M | 1 000 |
| | 1648 | 0 | D | T | 2 | a | U | Lorry | Front wall | M | 300 |
| Oct–Dec 1972 | 1649 | 0 | S | – | 2 | a | B | Crane | Side wall/floor | S | 2 000 |
| | 1658 | 0 | D | T | 2 | a | B | Lorry | Front wall | M | 300 |
| | 1667 | 0 | D | U | 2 | b | B | Lorry | Front wall | M | 750 |
| | 1729 | 0 | D | S | 2 | a | B | Bus | Front wall | M | 300 |
| | 1747 | 0 | H | – | U | a | B | Lorry | 1 wall | M | 400 |
| | 1750 | 3(S) | S/F | – | 2 | a | B | Lorry | 2 walls/floor/roof | S | 2 500 |
| | 1759 | 0 | D | D | 2 | b | B | Lorry | 2 walls/floor | S | 4 000 |
| | 1774 | 0 | S | – | 2 | a | B | Lorry | Front wall/floor/roof | S | 5 000 |
| | 1775 | 0 | D | U | U | U | U | Lorry | Side wall | M | 750 |

*(V) in vehicle; (B) in building; (S) in street.
†M moderate; S severe.

| Quarter | Incident no | No of fatal casualties* | Building(s) affected | | | | | Vehicle type | Extent of significant damage | Rating† | Estimated cost of repairs (£) |
|---|---|---|---|---|---|---|---|---|---|---|---|
| | | | Use | Type (house only) | No of storeys | Age | Structure | | | | |
| Oct– Dec 1972 (cont) | 1786 | 0 | F | – | 2 | a | B | Crane | Upper wall | M | 300 |
| | 1823 | 0 | S | – | U | a | B | Lorry | 1 wall | M | 500 |
| | 1837 | 0 | S/F | – | 2 | b | B | Lorry | 1 wall/floor | M | 600 |
| | 1885 | 0 | S | – | 2 | a | B | Car | Front wall | M | 600 |
| Jan– Mar 1973 | 1963 | 0 | D | U | 2 | b | B | Car | Front wall | M | 400 |
| | 1981 | 0 | D | U | U | a | B | Car | Front wall | M | 400 |
| | 2009 | 2(S) | F | – | U | b | B | Lorry | Front wall/int wall | M | 1 000 |
| | 2027 | 0 | D | D | 2 | a | T | Lorry | Front wall/roof | M | 1 500 |
| | 2044 | 0 | D | S | 2 | a | B | Lorry | Front wall | M | 750 |
| | 2071 | 0 | D | U | 2 | a | U | Lorry | Front wall | M | 500 |
| | 2072 | 0 | D | U | U | U | U | Lorry | Front wall | M | 400 |
| | 2105A | 0 | X | – | U | c | U | Bus | 1 wall/floor | M | 600 |
| | 2119 | 0 | D | U | U | a | U | Van | 1 wall | M | 400 |
| | 2212 | 0 | D | T | 2 | a | M | Lorry | Front wall | M | 500 |
| | 2240 | 0 | D | U | 2 | b | B | Lorry | Front wall | M | 500 |
| Second series | | | | | | | | | | | |
| April– June 1973 | 39 | 0 | D | U | U | a | U | Lorry | Front wall | M | 250 |
| | 115 | 0 | D | U | U | a/b | U | Lorry | Front wall | M | 2 500 |
| | 184 | 0 | 2D | T | 2 | a | B | Lorry | 2 front walls | M | 750 |
| | 226 | 0 | D | B | 1 | c | B | Taxi | 1 wall | M | 200 |
| | 239 | 0 | D | D | 2 | a | B | Lorry | 2 walls | M | 400 |
| July– Sept 1973 | 251 | 0 | S | – | U | a | B | Lorry | 1 wall | M | 450 |
| | 265 | 0 | S | – | U | a | U | Lorry | Front wall | M | 250 |
| | 280 | 0 | D | B | 1 | b/c | U | Lorry | Front wall | M | 200 |
| | 303 | 0 | D | S | 2 | b | U | Bus | Front wall/floor | M | 500 |
| | 370 | 0 | D | U | U | a/b | U | Lorry | Front wall | M | 250 |
| | 379 | 0 | S | – | 2 | a | B | Lorry | Front and side walls | M | 1 250 |
| | 385 | 0 | D | B | 1 | b | B | Bus | Front and side walls | M | 600 |
| | 407 | 0 | S | – | U | a | U | Car | Front wall | M | 650 |
| Oct– Dec 1973 | 421 | 0 | D | U | 2 | a | B | Lorry | Front wall | M | 400 |
| | 434 | 0 | 2S | – | 2 | a | B | Bulldozer | 2 front walls/int wall | M | 1 750 |
| | 470 | 0 | D | D | 2 | c | B | Van | Front and side walls | M | 1 750 |
| | 499 | 0 | D | U | U | U | U | Car | Front wall | M | 450 |
| | 507 | 0 | D | U | 2 | U | B | Tanker | 1 wall | M | 600 |

*(V) in vehicle; (B) in building; (S) in street.
†M moderate; S severe.

| Quarter | Incident no | No of fatal casualties* | Building(s) affected | | | | | Vehicle type | Extent of significant damage | Rating† | Estimated cost of repairs (£) |
|---|---|---|---|---|---|---|---|---|---|---|---|
| | | | Use | Type (house only) | No of storeys | Age | Structure | | | | |
| Jan– Mar 1974 | 678 | 0 | D | T | 2 | b | B | Car | Front wall | M | 450 |
| | 690 | 0 | D | T | 2 | a | B | Car | Front wall | M | 300 |
| April– June 1974 | 817 | 1(B) | S | – | 2 | a | B | Lorry | Front wall/floor | M | 1 750 |
| | 820 | 0 | D | S | 2 | c | B | Van | Side wall | M | 1 500 |
| | 840 | 0 | D | S | 2 | c | B | Lorry | Front wall | M | 350 |
| | 844 | 2(B) | 2F | – | U | b/c | U | Lorry | Front wall | M | 1 000 |
| | 847 | 0 | D | U | 2 | a | B | Lorry | 1 wall | M | 500 |
| | 858 | 0 | D | U | 2 | a | B | Lorry | 1 wall | M | 250 |
| | 963 | 0 | D | T | 2 | a | B | Bus | Front wall | M | 400 |
| July– Sept 1974 | 1021 | 0 | D | S | 2 | c | B | Lorry | Front wall | M | 600 |
| | 1053 | 0 | D | S | 2 | b | B | Crane | Side wall/roof | S | 2 500 |
| | 1100 | 0 | 3D | U | U | a | B | Fire engine | 3 front walls | M | 1 750 |
| | 1102 | 1(S) | S | – | 2 | a | B | Lorry | Front wall | M | 1 500 |
| | 1110 | 0 | D | U | U | a | U | Car | Front wall | M | 350 |
| | 1157 | 0 | D | U | U | a/b | B | Car | Front wall | M | 500 |
| | 1168 | 1(S) | S | – | 1 | c | B | Lorry | Front wall | M | 500 |
| Oct– Dec 1974 | 1220 | 0 | D | T | 2 | a | B | Car | Front wall | M | 500 |
| Jan– Mar 1975 | 1537 | 0 | D | S | 2 | a | B | Car | Side wall | M | 1 000 |
| | 1570 | 0 | D | U | 2 | b | B | Car | Front wall | M | 500 |
| | 1631 | 0 | D | T | 2 | a | B | Bus | Front wall | M | 1 500 |
| | 1678 | 0 | D | S | 2 | c | B | Dustcart | Corner | M | 750 |
| April– June 1975 | 1722 | 0 | D | U | 2 | a | U | Car | 1 wall | M | 1 500 |
| July– Sept 1975 | 2074 | 0 | E | – | 1 | U | B | Lorry | 1 sub-station | M | 3 000 |
| Oct– Dec 1975 | 2083 | 0 | 2D | T | 2 | a | B | Bus | 2 front walls | M | 1 000 |
| | 2257 | 0 | S | – | 1 | a/b | B | Car | Front wall | M | 1 500 |
| | 2264 | 0 | D | D | 2 | a | B | Lorry | End wall/floor/roof | S | 6 000 |
| | 2305 | 0 | S/F | – | 2 | a | M | Tanker | Front wall/floor | M | 3 000 |
| | 2333 | 0 | 2D | S | 2 | a | B | Lorry | 2 front walls/party wall | S | 3 000 |
| Jan– Mar 1976 | 2424 | 0 | D | U | U | a | B | Car | Front wall | M | 500 |
| | 2466 | 0 | D | T | 2 | a | B | Car | Front wall | M | 1 500 |

*(V) in vehicle; (B) in building; (S) in street.
†M moderate; S severe.

| Quarter | Incident no | No of fatal casualties* | Building(s) affected | | | | | Vehicle type | Extent of significant damage | Rating† | Estimated cost of repairs (£) |
|---|---|---|---|---|---|---|---|---|---|---|---|
| | | | Use | Type (house only) | No of storeys | Age | Structure | | | | |
| Jan– Mar 1976 (cont) | 2471 | 0 | S | – | 2 | a | U | Lorry | Front wall | M | 500 |
| | 2550 | 0 | W | – | 1 | b/c | B | Lorry | 1 wall | M | 1 000 |
| | 2571 | 0 | D | S | 2 | a | B | Lorry | Front wall | M | 1 000 |
| | 2573 | 0 | D | S | 2 | c | B | Lorry | Front wall | M | 750 |
| April– June 1976 | 2702 | 0 | D | U | U | U | U | Car | Front wall | M | 750 |
| | 2704 | 0 | H | – | U | a | U | Car | Front wall | M | 500 |
| | 2716 | 0 | D | B | 1 | b | M | Car | Front wall | M | 400 |
| | 2764 | 0 | D | B | 1 | a | B | Car | Front wall | M | 2 000 |
| | 2790 | 0 | D | T | 2 | a | B | Lorry | Front wall | M | 750 |
| | 2843 | 0 | D | S | 2 | c | B | Crane | Gable end wall | M | 400 |
| | 2907 | 0 | X | – | U | c | B | Lorry | 1 wall/roof | M | 1 000 |
| July– Sept 1976 | 2978 | 0 | S | – | 2 | a/b | B | Car | Side wall | M | 450 |
| | 2980 | 0 | D | D | 2 | c | B | Crane | 1 wall/roof | M | 1 000 |
| | 3014 | 0 | S | – | 1 | a | B | Lorry | Pub extension | M | 2 500 |
| | 3037 | 0 | D | S | 2 | a | B | Lorry | Front wall | M | 600 |
| | 3072 | 0 | D | S | 2 | a | B | Lorry | Outhouse | M | 700 |
| | 3105 | 0 | D | U | U | a/b | U | Van | Front wall | M | 500 |
| | 3148 | 0 | X | – | 1 | b | B | Lorry | Front wall | M | 1 000 |
| Oct– Dec 1976 | 3183 | 1(S) | D | S | 2 | a | B | Lorry | Front wall | M | 1 000 |
| | 3201 | 0 | D | U | 2 | a | M | Car | Front wall | M | 400 |
| | 3215 | 0 | D | U | 2 | c | B | Car | Garage wall | M | 500 |
| | 3231 | 0 | D | U | U | a | U | Car | Front wall | M | 750 |
| | 3266 | 0 | D | S | 2 | a | B | Lorry | Side wall | M | 750 |
| | 3329 | 0 | D | U | 2 | a | M | Lorry | Front wall | M | 2 000 |
| | 3366 | 0 | D | U | U | U | U | Van | Front wall | M | 750 |
| | 3377 | 0 | D | B | 1 | a | B | Car | Front wall | M | 2 000 |
| | 3412 | 0 | D | U | 2 | a | B | Car | Front wall | M | 450 |
| Jan– Mar 1977 | 3415 | 0 | D | D | 2 | a | M | Lorry | Front wall | M | 1 500 |
| | 3418 | 0 | D | B | 1 | c | B | Van | Front wall | M | 500 |
| | 3590 | 0 | D | S | 2 | a | B | Bus | Gable end wall/floor | S | 10 000 |
| | 3598 | 0 | D | U | 2 | a | B | Car | Front wall | M | 2 500 |
| | 3639 | 0 | D | S | U | a | B | Lorry | Outhouse | M | 400 |
| | 3698 | 0 | D | U | U | a | U | Lorry | Front wall | M | 500 |
| | 3715 | 0 | D | U | U | b/c | U | Car | Front wall | M | 500 |
| | 3732 | 1(V) | S | – | U | b | B | Car | Front wall | M | 750 |

*(V) in vehicle; (B) in building; (S) in street.
†M moderate; S severe.



Printed in England for Her Majesty's Stationery Office by
Hobbs the Printers of Southampton
(1206) Dd0586189 K12 7/78 G3927